WHY RECYCLE?

PROCEEDINGS OF THE RECYCLING COUNCIL ANNUAL SEMINAR
BIRMINGHAM / UNITED KINGDOM / 17 FEBRUARY 1994

WHY RECYCLE?

Edited by
A.K.M.RAINBOW
The Recycling Council, Bedlington, United Kingdom

A.A.BALKEMA / ROTTERDAM / BROOKFIELD / 1994

The texts of the various papers in this volume were set individually by typists under the supervision of each of the authors concerned.

Published by
A.A. Balkema, P.O. Box 1675, 3000 BR Rotterdam, Netherlands
A.A. Balkema Publishers, Old Post Road, Brookfield, VT 05036, USA

ISBN 90 5410 367 1

Printed in the Netherlands

Why Recycle?, Rainbow (ed.) © 1994 Balkema, Rotterdam, ISBN 90 5410 367 1

Table of contents

Why Recycle?, Rainbow (ed.) © 1994 Balkema, Rotterdam, ISBN 90 5410 367 1

Foreword

Eur. Ing. Dr A. K. M. Rainbow
Executive Chairman The Recycling Council

Bachelor of Science (Hons) (Civil Eng.)
Doctor of Philosophy (Mining Eng.)
Chartered Engineer
Fellow, Institution of Civil Engineers
Fellow, Institution of Mining & Metallurgy
Fellow, Institution of Highways & Transport
Fellow, Institution of Water & Env. Management
Fellow, Geological Society
Member, Geotechnical Society

Organiser & Editor

First, Second and Third International Symposium on the Reclamation, Treatment and Utilisation of Coal Mining Wastes

Why recycle?

Questions, only questions?

Lord Strathclyde when Minister responsible for environment and recycling recognised the importance that the recycling industry was able to speak to government through a representative body. It was considered that this ability would be of great help in developing future policy and legislation.

Therefore The Recycling Council consider that they have an obligation not only to their members but also to those organisations whether quasi or otherwise to demonstrate that recycling though highly desirable does not happen by will alone. In order to have a lasting effect the process must be commercially worthwhile and if these industries are not viable the inevitable question will be asked – then why recycle?

Lord Reay, Lord Strathclyde's predecessor speaking at the UKRC seminar; 'Recycling – A commercial perspective' enthusiastically endorsed the theme of the seminar stressing that 'sustainable recycling cannot take place unless it is based on sound economics'.

It would appear that it is only in recent years that attention has been focused on the need for a proper strategy embracing collection, sorting and utilisation of surplus materials which is more commonly referred to as recycling. Perhaps the industries concerned with recycling have preferred to operate without 'interference' from government secure in the knowledge that they were the experts and jealously guarded their commercial autonomy.

Does Britain have a strategic policy on recycling? and should it have? Is it advisable or even helpful for government to provide a legislative framework to control recycling policy? Can government legislate for leadership in waste minimisation and coerce a design for recycling philosophy? Under the German 'Dual System' (DSD) the manufacturers of packaging, appliances, cars etc., must take back their products at the end of their life-cycle. What the German legislators forgot to calculate was the viable size of processing capacity and available markets. This miscalculation is causing serious difficulties in countries where the surplus is being dumped. Will it take British legislation to rectify the situation? Is legislation a legitimate tool to alter market forces? Certainly these points will be addressed at this seminar.

An equally fair question is whether a seminar of this type can attempt to address the complex issues involved in ferrous and non-ferrous scrap, tyre recycling, paper, board, plastic, cans, glass, textile recycling industries or in sewage sludge, composting, domestic waste etc. etc.

The Recycling Council may have to consider a greater variety of complex issues under its Constitution since its newer members have shown that there are many aspects which Council have not debated as an example; in the debate on the reclamation of packaging, the practice of reconditioning for re-use is never brought into the equation. This oversight may have arisen because of the incorrect supposition that reconditioning of such 'packaging items' as aluminium, steel and plastic drums does not deserve consideration as a feasible option. This sector of the recycling industry is somewhat aggrieved since they would argue that they make a not inconsiderable contribution to the recycling activities of the Nation and expect more recognition. Through The Recycling Council they will now have an opportunity to express their views and of course that it is the ultimate purpose of The Recycling Council seminars – to give you an opportunity to voice your views.

Why Recycle?, Rainbow (ed.) © 1994 Balkema, Rotterdam, ISBN 90 5410 367 1

Introduction by Lord Ezra of Horsham, MBE

Lord Ezra of Horsham, MBE
President of The Recycling Council

This volume is a compilation of the views of many international experts on the need for recycling and presented by them on the occasion of the Wasteman Exhibition 1994. I can commend it as a major initiative of the newly-formed Recycling Council (TRC).

No one now doubts the importance of recycling as much waste material as possible for environmental reasons. However, there must be an economic justification and recycling will only be successful if the right conditions are created.

This can either be done by legislation and regulation or by influencing market forces. There has already been a good deal of legislation in this sector arising both in Britain and from Brussels. But it would be wrong for this to be carried too far and already some of the directives have created problems.

Influencing markets is by no means easy but is essential if progress is to be made. Such bodies as The Recycling Council help to fulfil this function. Objective study of recycling possibilities are continually needed in order to point to the best prospects in changing market situations. An excess provision of recyclable material in a weak market situation can cause considerable harm. On the other hand some markets can gain by the provision of additional recycling capability.

These are among the issues dealt with in the selection of papers.

It will be the task of The Recycling Council to give these matters continuous attention and I hope it will get the support of all those who are involved in any way in the problems and prospects for recycling. The Recycling Council is the successor to the United Kingdom Recycling Council and has been reconstituted on a wider basis as the preamble to the new constitution will show.

Why Recycle?, Rainbow (ed.) © 1994 Balkema, Rotterdam, ISBN 90 5410 367 1

Developments in the biology of compost processes

A.A. Keeling
Harper Adams Crop and Environment Research Centre, Newport, Shropshire, UK
J.A.J. Mullett
International Process Systems, Feltham, UK
I.K. Paton
Oadby, Leicester, UK

ABSTRACT: Composting is an ideal way to process organic wastes for use in agriculture and horticulture. This paper summarises recent developments in the understanding of the biology of composting and its applications.

1 INTRODUCTION

Composting can be defined as the aerobic, microbiological digestion of organic wastes to yield a safe useful product for application in agriculture and horticulture. Since 1990, when the Environmental Protection Act came into force in the UK, the government has set a target of 25% of domestic waste to be recycled by the year 2000. Local authorities are the organisations with the responsibility for achieving this target. Historically, advances in recycling have been hindered by cost and lack of markets for recycled products. The majority of authorities now recognise that the organic element of waste (65% by weight, including paper) is the most important fraction to be recycled effectively if the target is to be reached. There are other compelling reasons to process and reuse as much of the organic fraction of waste as possible, most notably, the fact that organic matter causes most of the pollution, landfill gas and leachate problems now experienced at an increasing number of landfill sites.

Composting has been practised the world over for many centuries and is now a key waste management procedure particularly in the USA and continental Europe, but surprisingly has been largely ignored in the UK. Problems with composting of waste have resulted from the presence of high quantities of contaminating materials in the products, especially glass shards, heavy metals and pathogenic organisms. However, a number of procedures have been set in place by both collection and disposal authorities which allow the effective segregation of organic waste at source. For example, organic waste may be collected relatively uncontaminated if households are provided with two bins, with one devoted solely to compostable material, and alternate fortnightly collection. In addition, many disposal authorities are investigating the possibility of segregating green (garden) waste at civic amenity sites and considerable degrees of success are being achieved. The market outlets for compost are being aided by the commitment of local authorities to use the products themselves should higher value markets prove to be unavailable.

Surprisingly, the biological and biochemical aspects of compost processes are still poorly understood and, to make matters worse, there are a considerable number of conflicting statements to be found in the literature which tend to discourage some people from composting. Such statements are: composts only have a minimal fertiliser value; unstable (biologically active) composts should not be used because they adversely affect plant growth and thirdly, paper products (lignocellulose) materials are not compostable. The aim of this article is to make it clear that none of the above statements can be justified from experimental observations.

2 UNSTABLE COMPOSTS HAVE VALUABLE FERTILISING AND NUTRIENT RELEASING PROPERTIES

"Unstable" composts are those which have not fully degraded and are colonised by actively growing microorganisms. Traditionally, it has been believed that unstable composts should not be used in agriculture or horticulture because of phytotoxic (plant toxic) effects. However, this only occurs in the initial stages of seed germination and growth and in media where the proportion of unstable compost is high. Experiments on amenity ryegrass (*Lolium perenne*) grown in media containing unstable refuse-derived compost (RDC) over the whole range of application rates (from 0 to 330 kg/m^3) show that up to 100 kg/m^3 no significant phytotoxic effects are observed (see fig 1). Grass grown in 100% RDC (330 kg/m^3) experienced significant early germination and growth inhibition but by 5 to 6 months of growth, such plants were growing strongly implying that they were obtaining significant concentrations of nutrients over an extended period. They also possessed extensive, healthy root systems.

Similarly, the effects of applying 20% unstable RDC to growing mixtures containing young hardy "*Primula*" hybrids were tested and are shown in table 1. Plants in media containing RDC were compared with those grown in media containing conventional N-P-K balanced fertiliser, scoring the plants on a 1-5 basis showed that media containing RDC consistently out-performed those containing the conventional fertilisers.

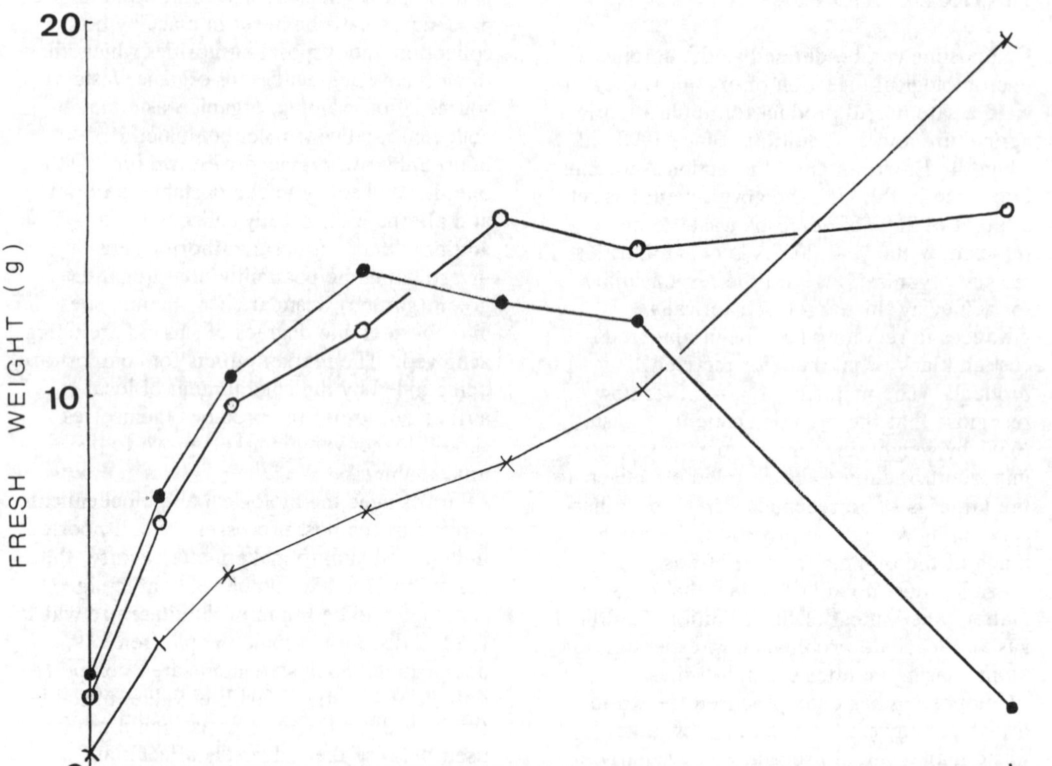

Figure 1 Fresh yields of perennial ryegrass in media containing unstable RDC. Harvests at 40 (●), 91 (o) and 180 (x) days after sowing.

Table 1 Primula growth trial scores, 1 = Poor, 5 = Very good

Growing Media	With Fertiliser	With 20% RDC	Growing media means
Peat	3.67	3.65	3.66
Peat/Bark	3.56	4.18	3.87
Coir	3.03	3.22	3.12
Bark	3.13	3.59	3.36
Treatment Means	3.34	3.66	

3 UNSTABLE COMPOSTS CONTAIN BOTH GROWTH INHIBITING AND GROWTH STIMULATING FACTORS

Traditionally, the phytotoxicity of composts has been determined using cress seed germination tests originally developed by Zucconi et al (1981) and recently modified by Keeling et al (in press 1994) to show the impact of different molecular weight fractions of compost on seed germination.

It has been demonstrated (Table 2) that while the low molecular weight components of unstable compost inhibit seed germination and growth, the high molecular weight fraction stimulates growth. As yet, the substances causing this growth stimulation are unidentified, but they may be responsible for the development of the unusually healthy root systems ultimately developed by plants in RDC media.

Table 2 Effect of high and low molecular weight fractions on the germination of cress in water extracts of RDC

Molecular size	Mean % germination enhancement (+) or inhibition (-)
Large (> 10 kD)	+ 21.9 (± 20.1)
Small	- 46.1 (± 16.7)

4 PAPER PRODUCTS CAN BE COMPOSTED (WITH ASSOCIATED ATMOSPHERIC NITROGEN FIXATION)

The composting of putrescible organic matter is thought to be dependent on the carbon to nitrogen ratio of the organic fraction which should ideally be below 30:1 to facilitate the process (Golueke and Dias, 1990) The N content needs to be high as microbes require it to synthesise proteins and nucleic acids. Therefore, lignocellulose wastes which contain very little nitrogen are highly resistant to biodegradation under normal conditions. The observation that lignocellulose waste can be converted into a nitrogenous compost without man's intervention is new and potentially highly valuable.

Densified refuse-derived fuel (d-RDF) is manufactured from mixed domestic refuse. In the first stage of production, the majority of putrescible matter and inert heavy material is removed by passage through a trommel screen containing 5cm apertures. The paper and plastic fraction is then further dried and shredded followed by densification and pelletisation at 100°C. The final product can be used as a fuel, either alone or mixed with coal. It has a typical moisture content of 4%, too low to sustain the proliferation of microorganisms (Golueke and Dias, 1990).

When the material is exposed to the elements for several months, moisture is absorbed and the interiors of the piles heat up and fungal growth can be seen. Temperatures can rise to as high as 80°C during composting. Biological and physico-

Table 3 Biological and physico-chemical characteristics of composting d-RDF and active RDC.

Parameter/Organism	Composting d-RDF	Fresh d-RDF	Active RDC
Sample temperature (°C)	80	Ambient	70-80
pH	6.88	6.99	6.82
Total N (Kjeldahl %)	1.35	0.5	1.95
Total bacteria x 10^6/g (25°C)	90	0.0014	208
Total bacteria x 10^6/g (55°C)	27	0.0009	26
Actinomycetes x 10^3/g (55°C)	380	0.19	6.3
Penicillium x 10^3/g (25°C)	70	ND*	ND
Mucor x 10^3/g (25°C)	50	ND	ND
" (37°C)	10	ND	ND
Mortierella x 10^3/g (25°C)	30	ND	ND
" (37°C)	4	ND	ND
Geotrichium x 10^3/g (25°C)	ND	ND	7.4
Yeasts x 10^3/g (25°C)	810	ND	> 3000
" (37°C)	450	ND	> 3000

(*ND : Not deleted)

chemical parameters of d-RDF at this stage of decomposition were determined according to standard MAFF/ADAS procedures, and compared with data for freshly synthesised d-RDF and active refuse-derived compost (RDC, derived from the screened putrescible fraction of domestic refuse). The results of the analyses are shown in Table 3. The temperature and thermophile content of composting d-RDF was essentially similar to the RDC, but pseudomonads predominated in composting d-RDF. The nitrogen content of this material was almost three times higher than fresh d-RDF, which never contains more than 0.7% nitrogen and has a typical carbon to nitrogen ratio of 94:1.

The conclusion from these observations is that an organism(s) predominating in composting d-RDF is responsible for the nitrogen accumulation and fixation. Organisms found in substantial numbers in composting d-RDF are pseudomonads, Actinomycetes, *Mucor*, *Penicillium* and *Mortierella*. Of these, only pseudomonads are believed to be capable of fixing nitrogen effectively, possibly as dinitrogen gas (N_2; Barraquio et al, 1988; Gibson et al, 1988). Nitrogen is apparently also effectively scavenged as NOx or ammonia from rainwater or the atmosphere by pseudomonads (Postgate, 1987)). Dinitrogen gas fixation in *Clostridium butyricum* has been observed by Chapman et al, 1992, who showed that under ideal conditions this species could fix 6 mg of N per g of straw decomposed. In the present system, similar quantities of N per g of material are fixed, though the conditions are not directly comparable.

These observations have significance for the biological processing of high C, low N wastes. Clearly, composting can occur under favourable ambient conditions and nitrogen fixed as a nutrient source for bio-oxidation.

This may be capitalised upon in the production of nitrogenous organic fertilizers and soil conditioners. Further work is necessary to identify precisely the conditions and organisms involved in the nitrogen fixation process.

CONCLUSIONS

It is now accepted that composting will take its place as a key waste management process in the UK. However, the products and processes still need to be evaluated carefully (and scientifically) to enable the most effective applications to be realised.

ACKNOWLEDGEMENTS

We thank Dr. M.H.B.Hayes for valuable advice and Miss F. Pickering for preparing the typescript.

REFERENCES

Barraquio, W.L., Dumont, A. and Knowles, R. (1988) A method for counting free-living aerobic N_2-fixing bacteria. in: Nitrogen fixation: Hundred years after. Proceedings of the 7th International congress on N: Nitrogen fixation. Gustav Fischer (Stuttgart/New York), 790.

Chapman, S.J., Veal, D.A. and Lynch, J.M. (1992) Effect of oxygen concentration on dinitrogen fixation and volatile fatty acid production by *Clostridium butyricum* growing in association with fungi on cellulose and wheat straw. J. Appl. Bacteriology, 72, 9-15.

Gibson, A.H., Halsall, D.M. and Roper, M.M. (1988) Nitrogen fixation associated with residue breakdown. In: Nitrogen fixation: Hundred years after. Proceedings of the 7th International Congress on N: Nitrogen fixation. Gustav Fischer (Stuttgart/New York), 753-758.

Golueke, C.G., and Dias, L.F. (1992) Understanding the basics of composting. Biocycle, 56-59.

Postgate, J. (1987) Nitrogen Fixation, 41. Edward Arnold.

Zucconi, F., Forte, M. Monaco, A. and de Bertholdi, M. (1981). Biological evaluation of compost maturity. Biocycle 22, 27-29.

Why Recycle?, Rainbow (ed.) © 1994 Balkema, Rotterdam, ISBN 90 5410 367 1

Market demand for MSW and sludge derived composts

L. R. Kuhlman
Universal Composting Corporation Ltd, Colorado, USA

ABSTRACT: Experiments have shown positive growth and yield results from the application of MSW and sewage sludge derived composts.

Data provided indicates that uses for compost exceed expected production. Revenues may be low, however the composting project is successful when the compost is utilized in a productive manner. The markets or outlets are known and the challenge is to develop the market.

Composting facilities are increasing rapidly in number throughout the U.S. and rapid expansion is expected in the U.K. The compost produced must be disposed of to appropriate outlets in order to ensure that composting is a practical waste management option. The outlets for municipal solid waste and/or sludge derived composts are generally recognized to be reclamation, silviculture, agriculture and horticulture.

DISCUSSION

For the purposes of this paper it is assumed that the pathogen level and heavy metal content of the composts meet existing and/or proposed standards. Markets depend upon compost quality and consistency of the product. MSW derived composts may not meet all of the standards for other composts and thereby limit their use. Failure of MSW composts to meet the highest quality and nutrient standards need not prevent their use in appropriate applications. Further processing may be required to satisfy certain end uses and/or to upgrade the compost for higher value markets.

An often expressed concern is whether or not the production of large amounts of compost will saturate the market causing problems of disposal. In order to determine whether or not the total compost produced from all sources could theoretically be utilized effectively, one must estimate both the production and the available market. Table 1 provides estimates of composts which are expected to be produced from the various available wastes in the U.S. (Batelle, 1992).

In the long term, MSW compost is predicted to be the major source of compost in the U.S. Significantly greater quantities of compost are currently being produced from green wastes than from any other waste. The U.S. EPA estimates that sewage treatment plants in the United States produce about 7.7 million tons (dry weight) of sewage sludge annually. Current estimates are that 15% of the sludge will be composted, giving rise to the potential compost production of 3 million tons. Amendments (bulking agents) such as wood chips, sawdust, straw and leaves are included in this figure. Decreases in land application, which is the disposal route of 50% of the sewage sludge increases compost production and exceed current estimates. This is due to increased concerns of nitrate leaching, run off, additional regulations and higher total costs.

A similar situation exists in the U.K. as shown in Table 2 (Paul, 1993). This table differs from Table 1 in that an estimate of the

Table 1. Potential and Current Compost Supply (U.S.A.)

Compost Source	Potential Compost (million tons)	Current Compost (million tons)
MSW	30	1
Sewage Sludge	3	2
Green Waste	15	5
Agriculture Waste	3	0.3
Total	51	8.3

Battelle, 1992

Table 2. Estimates of Organic Wastes (U.K.)

Arising	Million Tonnes	Compostable Est %	Est MT
Municipal Solid Waste	13	60	7.8
Sewage Sludge (20% D.M.)	6	100	6.0
Civic Amenity Site	5	80	4.0
Collected Commercial	3	70	2.1
Delivered Commercial	12	50	6.0
Industrial	68	5	3.4
Agricultural	80	90	72.0
Totals	187	54	101.3

Paul, 1993

total tonnage of the various raw wastes are given as a base point. The compostable fraction is expressed providing an estimate of the tonnage of waste which would be available for composting. If all wastes were composted, about 50 million tons of compost would be produced assuming 50% loss of weight. Utilization of other technologies available for MSW and sludge disposal will further reduce the expected output of compost to 20 to 25 million tonnes. Since the greatest percentage of the organic waste in the U.K. is agricultural and because much of this waste will, for the near future, be applied as raw waste, the maximum total production of compost may be nearer to 5 to 10 million tonnes. Calculations such as these are risky as assumptions are required as to future programs.

The potential U.S. demand for composts manufactured from various categories of wastes are shown in Table 3 (Batelle, 1992). Only bagged product sold through retail outlets has market penetration approaching the maximum. Compost used by nurseries in containers has excellent market penetration, but all other applications have low current penetration. When compared to Table 1, it is interesting to note that the practical applications for compost are an order of magnitude greater than the amount of compost that is predicted to be produced which in turn indicates that composts should not need to accumulate nor be landfilled.

Low current market penetration in silviculture, sod production, field nurseries, agriculture and final landfill cover indicate that there is great potential to increase the usage of MSW and sludge derived composts, as it is these composts that are likely to be used in the aforementioned categories.

The outlets in the U.K. are shown in Table 4 (Paul, 1993). On a theoretical basis, the U.K. has sufficient agricultural land to accommodate all of the composts produced. A practical consideration is that composts produced in metropolitan areas incur added transportation costs to reach agricultural land. In addition, there is an undetermined requirement for landfill final cover and other areas are severely in need or reclamation. Compost can replace a

Table 3. Potential Demand Summary (U.S.)

Application	Potential Compost Demand (million m^3)	Current Penetration (percent)
Landscapers	1.53	<20
Delivered Topsoil	2.84	<5
Bagged/Retail	6.15	80
Landfill Final Cover	0.5	<5
Surface Mine Reclamation	0.15	<5
Container Nurseries	.7	<50
Field Nurseries	3.1	<1
Sod Production	15.4	<1
Silviculture	80	<1
Agriculture	688	<1
Total Potential	798.4	<2

Battelle, 1992

significant amount of peat as well as lessor amounts of chemical fertiliser. Rates of application on agricultural land generally range from 5 to 20 tonnes per hectare compared to 50 to 1000 tonnes per hectare for reclamation.

Table 4. Market Place (U.K.)

* Agricultural Land	Million Acres
Forest & Woodland	5
Arable	17
Pasture	29
* Peat Replacement 3 Million M^3 pa	
* Fertiliser & Lime Over 700 Million pa	

Paul, 1993

The conclusion from the data is that there are sufficient beneficial outlets to provide for the disposal of the composts produced in both the U.S. and U.K. The task at hand then becomes one of market development and implementation of programs.

The primary value of composts are improvement of soil fertility and increased crop yield. Research has shown that both MSW and sludge derived composts improve soil fertility, improve plant growth and crop yield. Shiralipour (1992) presents a focused review of MSW composts used for agronomic, horticultural, silvicultural and reclamation purposes in *Biomass and Bioenergy*, Vol. 3, Numbers 3-4, pages 267-279.

The question of economics is one of paramount importance. Philosophies vary concerning the pricing of composts. Some enthusiasts insist that the compost must produce revenue. Others maintain that the compost must be distributed to a useful end use and any revenue received is a bonus providing a cost reduction to the composting endeavor. Many, including this author, believe that when composting is the lowest cost and/or best management option, successful distribution of compost ensures a successful program. Distribution programs may include give away and/or sale of compost at very low prices. Market development and education will increase demand and raise prices and this is a worthy goal. Certain expenditures are required for the promotion of the compost and a good eye catching easily remembered name should be employed.

Historically, farmers in both the U.S. and U.K. have received sludge at little or no cost and perhaps even paid a small fee. This has set a precedent for sludge and MSW composts. MSW composts are usually viewed as lower quality than sludge composts due to the contaminants of glass and plastic and lower nutrients. Further processing will improve the aesthetic qualities and help to reduce the opposition to MSW derived composts.

In the U.S., composts made from yard wastes (leaves, grass clippings and brush) are highly regarded and generally are sold for $5 to

$20 per ton at the compost site. A tiered pricing structure is common so that the purchaser of large amounts of compost pays lower prices. Some municipalities opt for give away programs. Revenue from green waste compost is expected to continue to be higher than from any other source.

In the U.S., composts from animal manures are routinely sold at $5 to $10 per ton at the compost site. The compost is used almost exclusively for agricultural purposes within a radius of 25 miles of the compost site. The compost sites are located on or very near the site of the waste generator (poultry, dairy, cattle feedyard). Composts from agricultural wastes constitute only 10% of the estimated potential production. (Table 1, Battelle, 1991).

MSW and sludge composts vary in price. Sludge composts in New England generally range in price from zero to less than $3.00 per cubic yard (Lang, 1990). Marketing at most facilities remains low profile and is dependent primarily on word of mouth.

Collins (1991) reports on research conducted on MSW composts that indicate an investment in compost for corn growing could be economically feasible. Application rates of 40, 250 and 1000 tons of MSW compost per acre produced substantially higher yields than the control. The total levels of compost were applied over a 5 year period but the effect of the compost was apparent for the following several years. Supplemental fertiliser was applied in all treatments. Application and delivery costs, as well as application rate, were found to influence the value.

Eight tons of compost applied per acre for five successive years provided calculated values for the compost of $11.00, $9.50, $4.50 and $1.00 per ton when application costs were $0.50, $2.00, $5.00 and $8.00 per ton respectively (Collins 1991). At 500 and 1000 tons per acre, value for the compost became negative at application rates of $1.75 and $1.00 respectively. These data indicate that the increase in corn yields from the application of 8 ton of compost per acre was economical. Residual value of the compost was not included. Assumption was made that all standards are met. Other market pressures such as farmer resistance to the product could (and would) likely bring the actual selling price to lower levels.

In the U.S. fertility programs for the growing of crops combine commercial fertiliser and compost unless the crop is grown under organic production guidelines. Levels of fertiliser were not studied in the research by Collins (1991). Relatively low levels of nitrogen (for corn) were applied. It is quite possible that the application of higher levels of fertiliser (especially nitrogen) may have changed the economics and lowered the apparent value of the compost.

There is great need for research to study the response of agronomic and horticultural plants to various levels of compost application and to determine optimum combinations of fertiliser and compost.

In Table 5, typical users and selling prices of sludge composts from 10 U.S. locations have been summarized by Donovan (1990). Prices shown are those received by the producer of the compost. Prices to the final end user are higher due to transportation and mark up. The amount purchased and the distance of delivery substantially influenced the price. It is notable that there are no prices of zero but there is one instance of one city paying for the removal of the compost.

Washington, D.C. has a highly successful program. It is interesting to note that Washington, D.C. is one of the larger producers of compost and the compost also commands one of the higher prices. This author was involved in the composting operations at this location at the inception of composting in 1979 and can verify that initially the compost was offered at give away prices. Because of education, promotion and distributorship network, the selling price had risen to £5 per cubic meter by 1990. This illustrates a general trend in the U.S. and similar results are likely in the U.K.

Table 5. Marketing Data for Sludge Composts (U.S.A.)

Location	Volume (m^3pa)	Method of Distribution	Typical Users	Selling Price by Owner (£/m^3)
Akron, OH	24,600	Service Company	Dept. Nat. Resources 70% Landscapers Topsoil Nursery	11 (<45m^3) 3.35 (45-150m^3) 2.60 (>150m^3)
Clayton Co. GA	3,800	Broker	Landscapers 75% Nursery 10%	4
Denver, CO	15,400	Direct Sale & Broker	Farmland 50% Gen. Public 40%	5.30 Dlvd. to 20 miles 10.60 Dlvd. to 100 mi.
Oakland, CA	19,200	Brokers	Landscapers	5.30 (1988)
Hampton Roads, VA	8,460	Direct Sale	Landscapers 40% Municipal 30% Gen. Public 20%	8 (<8m^3) 5.30 (8-385m^3) 4 (>385m^3)
Los Angeles Co., CA	210,000	Wholesale Supply Co.	Gen. Public & Nurseries	Contract
Nashville, TN	6,900	Wholesaler	Landscapers & Nurseries	3.35
Sarasota, FL	4,600	Wholesaler	Agriculture Right of Way	City Pays for Removal
Scranton, PA	11,500	Direct Sale & Broker	Landscapers 25% Municipal Give Away 25%	2
Washington, D.C.	30,750	Maryland Environmental Service	Landscapers 25% Gen. Public 40% Topsoil Dealers 25%	5

Donovan, 1990

CONCLUSION

Composting of sewage sludge and MSW are established in the U.S. and expect to be established in the U.K. as a practical, low cost method to replace ocean dumping (and to some extent, land application) of sludge and landfilling of MSW. Compost is a resource. The end result of the composting project is to dispose of the compost for a useful purpose, such as the growing of crops, trees or reclamation. Although one hopes to receive revenue from the sale of compost, projects can be truly successful with compost give away programs. Key elements in a program to increase revenue are promotion and education.

REFERENCES

Battelle and Albrecht Associates. Potential U.S. Applications for Compost. Report available from The Composting Council, 114 South Pitt Street, Alexandria, VA 22314.

Collins, A.R. 1991. How Much Can Farmers Pay for MSW Compost? BioCycle, 32 (10):66-69.

Donovan, J.R. 1990. Markets for Sludge Composts. BioCycle 31 (2):44-45.

Paul, P.J. 1993. Personal Communication.

Shiralipour, A., D. B. McConnell and W. H. Smith 1992. Uses and Benefits of MSW Compost: A Review and Assessment. Biomass and Bioenergy 3:3-4, pp 267-280.

Why Recycle?, Rainbow (ed.) © 1994 Balkema, Rotterdam, ISBN 90 5410 367 1

Composting of food processing wastes

R.B. Brinton
Woods End Research Laboratory, Stroud, UK
W.F. Brinton
Woods End Research Laboratory, Mt. Vernon, Maine, USA

ABSTRACT: Food processing wastes from a number of different sources have been successfully composted in projects in North America and Europe, set up and monitored by Woods End Research Laboratory, a composting consultancy firm. Types of wastes dealt with include coffee grounds, fruit pulp, offal, and vegetable processing wastes. This report looks at two major composting projects involving potato processing wastes, including potato processing slurry, reject fries, potato peels, and diseased potatoes. Problems were identified with potato slurry, including high moisture, low pH and levels of various compounds which had to be controlled due to adverse influences on the composting process. Recipes were found which not only compensated for such conditions but also provided an optimal mix of materials for composting. Careful mix-ratio formulation and project management avoided any odour or leaching problems. Products were pathogen-free and of good quality. Similar processes are applicable to other food processing wastes.

1. THE PROBLEM OF FOOD WASTE DISPOSAL

Food processors are facing rising costs and increasing difficulties in disposing of their wastes, owing to a significant extent to decreasing landfill space. Because of their high water and organic matter contents, the waste materials are not considered desirable for landfilling, with several European countries and a number of states in the USA limiting or banning such disposal altogether. The vegetable materials are also not desirable for incineration owing to their high moisture content and low calorific value.

As a consequence of difficulties faced, alternative options are often being considered by processors or waste disposal authorities. The nutrient content and the high organic matter content suggests a process which might also benefit agriculture or horticulture. In this way, substances would be 'recycled' and returned to the land.

Three routes are currently most used: aerobic degradation (composting), anaerobic digestion, and direct land application. The latter is most common when dealing with crop residues before processing; there are often severe limitations and disadvantages with direct land application after processing, and indeed also with some unprocessed vegetable wastes: the susceptibility to leaching (= water pollution and loss of nutrients), the presence of undesirable compounds (e.g. organic acids) and sometimes the presence of pathogens.

1.1. The Composting Option

If properly carried out, composting can deal with the difficulties encountered with raw food processing wastes. Organic materials are stabilised in the form of humus, making them less susceptible to leaching problems, undesirable compounds are broken down, and pathogens are eliminated through the high temperatures and microbial competition which occur during composting. In addition, a finished compost can be stored and used later. It can, naturally, also be sold to recover costs involved with processing.

For waste products to compost properly, the right ratios of nitrogen, carbon, water and air are required to generate the correct chemical reactions and provide the necessary conditions for microorganisms. For the small garden compost pile, the home gardener is not as critical about long composting terms or other seemingly minor problems, such as small whiffs of odour from a not so good mix. However, if the amount of the material is increased significantly, a thorough knowledge of the right conditions for good composting becomes an absolute necessity, particularly with food processing wastes or other wet and putrescible materials. With proper mixes of ingredients and a well set up and monitored process, the food processing wastes as well as the most varied of other biological waste materials can be turned into rich, quality composts.

2. COMPOSTING OF POTATO PROCESSING WASTES

As composting consultants for the food industry and municipalities, Woods End's project in composting quantities of potato culls and processing wastes in Maine in 1989 was the first of its kind in the USA. Following the success of this method of waste treatment, WERL participated in setting up larger scale potato waste composting projects for the Canadian Department of Agriculture on Prince Edward Island, with over 30 miles of windrows.

2.1. Disposal Practices

A major potato producing state in the USA, five to ten percent of the potato crops in Maine, on average, are discarded each year because of being undersized or diseased. Some of the healthy discards can be used for seed, feed, manufacture of starch or sold to secondary outlets for soups, potato puffs and other products. Others must be disposed of, either trenched, landspread or landfilled.

All three of the latter options posed problems. Trenching posed long-term problems with leaching, with buried materials taking a very long time (often years) to fully decompose. With many of the discards and culls being diseased, landspreading had severe limitations, putting restrictions on the use of the land for up to two to three years following application. Landfilling was becoming an expensive option for farmers, with landfill disposal costs ranging from $50 to $ 100 per ton.

With all of the disposal options posing potential environmental problems, the state Department of Agriculture saw that composting could represent the most sensible option, bringing a nutrient-rich waste product back to the land, if it could be shown to be effective in degradation and elimination of pathogens.

In addition, various potato processing industries in the state also have wastes, in the form of potato processing cake (slurry), chopped potato wastes, reject fries and potato peels as well as other forms. In the Maine composting project, it was decided to use processing sludge, which was a mixture of several types of processing wastes together.

2.2. Preliminary Studies

As a first step, different potato wastes were analysed to evaluate characteristics and suitability for composting. Various parameters were tested for, with representative results for potato culls and processing sludge shown in Table 1.

Table 1. Laboratory analysis of Potato Culls and Potato Processing Sludge

Sample Type	pH 1:1	Organic Matter %	C:N ratio	Moisture content %	Density kg/m^3
Potato culls	6.60	92.1	18.1	77.6	913.7
Potato sludge	3.80	91.1	28.1	74.7	931.5

Both the potato processing sludge and the culls exhibited a number of undesirable traits for composting, and complementary materials had to be found to compensate for the high density and moisture, high levels of volatile organic acids and low pH. Correction was also needed for C:N ratios of the culls. Materials analysed for complementing potato wastes included paper fibre sludge, sawdust, straw, and horse litter. Because of the close proximity of forestry operations and paper mills, sawdust and paper fibre sludge were chosen for the final operation. Analyses for the sawdust and paper fibre sludge used are shown in Table 2 below.

Table 2. Laboratory analysis of Softwood Sawdust and Paper Fibre Sludge

Sample Type	pH 1:1	Organic Matter %	C:N ratio	Moisture content %	Density kg/m^3
Sawdust	4.80	96.1	278.7	45.0	267
Paper sludge	8.46	90.4	250.2	65.6	676

The very low pH of some potato wastes required particular attention. Potato starch decomposes readily and under conditions of reduced aeration, as in processing sludge, will produce fermentative intermediary compounds. The acidity results from conversion of carbohydrates to organic acids, which not only are odorous but also phytotoxic as well. The low pH is also not amenable to rapid biological composting. Preliminary work investigated the effects of several additives for neutralising the acidity. Wood ash was decided on, in proportions which improved rate of CO_2 respiration. Too high additions inhibited respiration.

The determination of proper 'recipes' or mix-ratios

for raw ingredients, adjusting for parameters as discussed above, is of paramount importance for the establishment of optimal conditions for the composting process which follows. Problems can arise if attention is not paid to this, including with odours, leaching, phytotoxicity, low heat-rise and insufficient control of pathogens.

Woods End uses mix-ratio algorithms and other formulas for arriving at an initial recipe. Often a proposed mix is trialed first in laboratory bench-scale testing before moving to field scale projects. This provides for further fine-tuning and for predicting the course of composting in advance.

2.3. Set-Up of Compost Windrows

Six different recipes were established for field experimentation, using different proportions of ingredients. The various waste materials were then put into long windrows, and mixed with a p.t.o. driven compost windrow turning machine.

A special test pile was set up to evaluate pathogen survival. Potato pieces infected with pathogens including *Rhizoctonia solani* (Black Scurf), *Fusarium solani* (Dry rot), *Erwinia carotovora* (Soft Rot), and *Corynebacterium sepedonicum* (Ring Rot) were introduced by means of nylon sacks into a test pile and tracked during the composting for their survival. These were compared with uncomposted wastes and with other non-inoculated compost windrows.

The schedule for turning of the windrows was once per day for the first two weeks, and afterwards every 2-3 days. The piles were monitored for temperature, moisture and pH during the composting term.

2.4. Results

Four of the mixes performed exceptionally well, with temperatures over 55°C for more than 35 days. Temperature curves for two of these windrow mixes are shown in Figure 1. In two other windrow mixes, the compost piles failed to generate enough heat until further pH compensation was made, at the nine day point, after which the temperatures increased to over the 55°C point and performed similarly to the above two. Another windrow demonstrated that an alternative method of pH control with regular limestone was not as effective.

Our results indicate that the texture of the compost mix was an important factor in its success, and that phytotoxic traits in the raw potato materials were eliminated through the composting process. Frequent turning of the pile was needed to achieve particle size reduction for the effective elimination of bacterial and fungal pathogens. Figure 2 shows phytotoxicity bio-assay results with raw vs. composted potato culls.

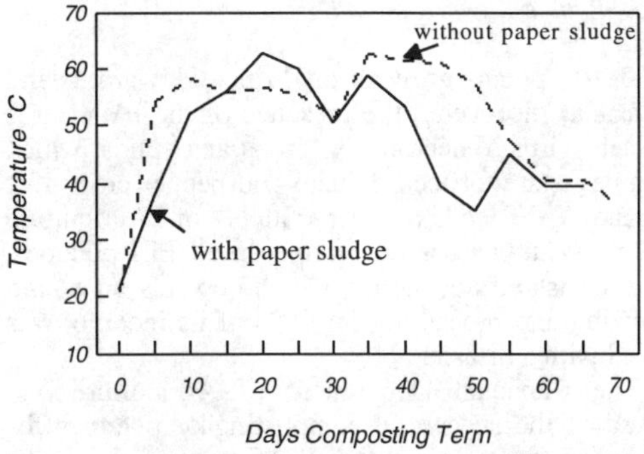

Fig. 1 Temperature vs. Time for Windrow Composting of Potato Culls and Processing Sludge, with and without Paper Fibre Sludge

The final composts showed an average total nitrogen-phosphorus-potash analysis (dry basis) of approximately 1-0.5-1, which is close to values expected for a composted dairy manure. The organic matter level averaged 35%, and salts were moderately low at 1.9 *mmhos/cm*.

Follow up field trials with the compost showed not only increased yields of potato and alfalfa with compost application compared to controls but also no disease carryover. (A complete presentation and discussion of set-up and results is available as separate publication from Woods End Research Laboratory.)

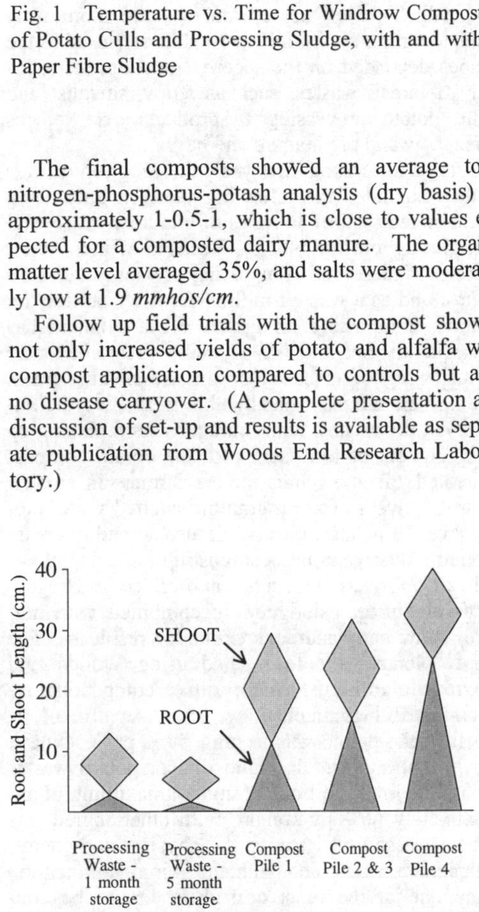

Fig. 2 Phytotoxicity Analyses of Raw and Composted Potato Wastes - Comparisons of Root and Shoot Length

Composting of potato culls and other wastes continues to be used in Maine. The University of Maine and the Maine Potato Board recommended it to growers for an estimated 10,000 tons of waste potatoes in 1993.

2.5. Prince Edward Island

In 1991, potato growers on Prince Edward Island (Canada) discovered the presence of the PVY virus (potato virus Y necrotic) in crops and spuds, which can devastate tobacco, tomato and pepper crops if it spreads. The U.S. Department of Agriculture promptly imposed a quarantine on all PEI potatoes, and the island's economy, which depends on potato growing and processing for 70% of its income, was faced with a disaster.

The Canadian Department of Agriculture had heard of the successful composting of potatoes for pathogen control in Maine and contacted Woods End, which helped set up and provide monitoring and testing for six initial sites. Materials used in the recipes depended on the specific location, using locally obtained wastes, such as straw, sawdust and some potato processing by-products, as well as horse, cow and pig manure and hay.

After a successful first year, the programme was further expanded the following year, encompassing an estimated total of 30 miles of windrows at several sites throughout the island. Because of the overall success of these PEI composting projects in disease control and as a waste handling-fertiliser production method, composting on Prince Edward Island has subsequently become the officially prescribed way of disposing of diseased or discard potato spuds. The farmers are also thereby able to recover the nutrients in the wastes, with some farmers replacing up to 20 percent of fertilisers with the potato compost generated. It is a bonus for the farmers in erosion control as well. The programme is cited with helping save the potato industry. It also served to create another industry: potato composting.

In a US Government sponsored project in Oregon, spuds and potato sludge were combined with manure, straw and hazardous explosive residues. The proper blends were formulated using Woods End mix-ratio algorithms, with positive composting results - for all ingredients.

All of the potato waste composting projects have shown that successful composting of potato wastes is possible in proportions of up to a maximum of approximately 60% by weight, with other ingredients being necessary to improve the physical and chemical characteristics, and with the initial mix criteria being one of the most critical factors to be considered.

3. CONCLUSION

The potato waste composting projects in Maine and Prince Edward Island have demonstrated that large scale composting is a viable means of disposal of excess potato cull and processing sludge waste which occur as a result of normal production operations. The inclusion of other waste streams not only compliments and aids the composting of the food wastes but also has the potential to alleviate disposal difficulties in other industries as well.

The findings of the projects showed that the success of the composting process was dependent on careful preparatory steps. Criteria for success include satisfactory heating, destruction of phytotoxic traits of the raw potato wastes, the elimination of disease substrate, and a compost product which otherwise fulfils quality standards and criteria.

Our research and field experiences with composting of other food processing wastes in North America and Europe, including coffee grounds, fruit pulp, offal, animal carcasses, fish processing wastes, and vegetable processing wastes, have shown similar results: Food wastes in themselves can have very one sided traits with regards to composting. Careful preparatory work, however, with determination of complimentary materials and mix-ratios, along with necessary quality control monitoring, can assure the composter of a successful outcome and a valuable, quality product.

Why Recycle?, Rainbow (ed.) © 1994 Balkema, Rotterdam, ISBN 90 5410 367 1

Cost effective recycling of domestic refuse by modified landfilling

T.Cairney
W.A. Fairhurst & Partners, Liverpool, UK

ABSTRACT: Despite public support and political will, little improvement has yet been achieved in recycling domestic wastes. This largely is due to the difficulties of obtaining public sector investment in a particularly severe recession. Revised (bioreactor) landfilling, practices, however, offer a solution which calls for no public funding and which produces significantly large energy resources.

1 INTRODUCTION

Volumes of domestic refuse in Europe have increased, on average, by 60% over the last 30 years, and no sign is yet apparent that this trend is reducing (ref.1). Quite obviously this current "waste culture" cannot be afforded indefinitely.

The political will for change is evident. For several years, the European Commission has had the problem high on its agenda. Draft Directives are in the late stages of negotiation, recycling and resource recovery targets are close to agreement, and most member states have devised recycling plans. Yet so far little or no improvement is apparent - European domestic wastes still predominantly are lost into traditional landfills.

A lack of public support cannot be seen as a cause of this lack of progress. Even when home occupiers have been asked to hand sort their refuse (a policy which now is seen to be less than fully effective) the levels of co-operation gained invariably have proved to be extremely high.

The problem, quite simply, has been the failure to invest in adequate recycling facilities. The best publicised case - that of the German "Green dot" (D.S.D.) system - makes the point. This recycling authority succeeded in persuading both German industry and public to sort refuse, and then began to receive vast quantities of wastes for treatment. However, cost effective recycling systems were not always available, and so the D.S.D. simply chose to export a large part of the problem to whatever neighbouring states would receive its wastes. The effects - of course - were to disrupt the indigenous recycling markets of the importing states. The UK, for example, received a 480% increase in German waste plastics in 1993 (often imported either at zero cost or with a subsidy inducement) and as a result the economics of its own plastics market were badly harmed. France, which also took 800,000 tonnes of German wastes in 1992, had the same experience with its paper reclamation system (ref.2). Had the Germans established their own recovery systems, before commencing the "Green dot" system, these problems could have been avoided. However, this obviously sensible approach calls for large scale investment, and that - in a Community wide economic recession - is more and more difficult to afford. Even quite simple composting approaches for domestic refuse can prove surprising expensive to set up (ref.3), and funds simply are not available in a period when state support for recycling initiatives is to be reduced (ref.4) and when the prices being paid to local authorities who collect economically recyclable wastes are falling.

There are few imponderable technical problems in recycling, political will for recycling exists, and the policy is apparently supported by a majority of the public, but investment remains the major hurdle. Since the economic situation looks unlikely to improve for some years, it is reasonable to examine reclamation options which can be utilised without any need for public financing.

2 TRADITIONAL LANDFILL

A variant of traditional landfill practices offers a solution for recovering resources from domestic refuse, calls for no investment funds, and produces significantly high energy resources.

Suggesting landfilling as a recycling option will appear strange.

Landfill is the least favoured waste disposal choice (ref.5) not only because of inevitable loss of recoverable resources that traditional practices imply, but because landfills give rise to serious and long term environmental degradations. Landfill gases, which will be given off for 30 to 50 years after such a landfill has closed, have serious effects on the atmospheric "greenhouse" condition, and pose significant risks to nearby home occupiers and to plant life. Similarly, most existing landfills generate polluting leachates for an equally long time, and these are known to be a major background cause of poor water qualities in rivers and aquifers.

These adverse features, however, are <u>not</u> inevitable consequences of landfilling, but arise directly from the operational practices and imperatives which govern today's landfills.

Landfill operators currently have as their prime aims to pack as much waste as possible into whatever storage volume is available, and to seek to do so as quickly and as cheaply as regulations permit. Thus a normal landfill is typified by:

1 A very small resident staff. As few as 7 employees can often be on site, even on a large (5 x 10^6 cubic metre) landfill.
2 An emphasis on using very large waste cells (perhaps 2ha or more in plan area) into which wastes are packed as tightly as possible, and then compacted.
3 A relative lack of interest in those factors (e.g. moisture content, material type, insulation etc.) which govern the rate and progress of biodegradation.

Because of these factors, degradation is slow,

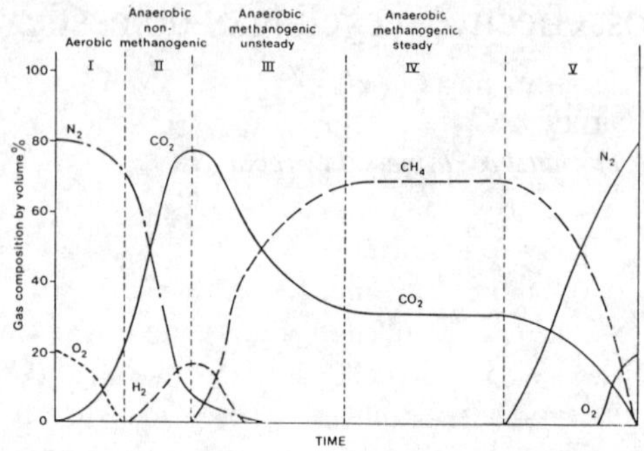

Figure 1 Phases in landfill gas generation (schematic)

variable, and tends to operate in the unstable methanogenic regime (phase III - fig.1). Thus long term environmental nuisances are created, both gases and leachates are formed, and a landfill site will seldom become stabilised (and available for any future re-use) for up to 50 years.

3 REVISED (BIOREACTOR) LANDFILL PRACTICES

To avoid the adverse features noted above, the critical starting point is to examine what conditions are necessary for complete and ultra rapid decomposition, and then determine whether or not the decomposition outcomes will be as required.

There is, in fact, no difficulty in establishing these most effective degradation conditions. Research findings (e.g. refs. 6 & 7) have been available for many years to show that if conditions in a landfill are made especially favourable for the growth of appropriate bacterial populations then rapid degradation occurs in a stable methanogenic state (phase IV - fig.1). Landfill gas outputs are maximised, leachates tend not to be generated, fatty acid contents decline and the time needed before the surface above a landfill becomes stable and free from subsidence is much reduced.

These environmental conditions, most conducive to rapid bacterial development, simply, are:

1 To ensure that the landfill material is relatively uniform and as consistently degradeable as possible. The presence of

non-degradeable materials (metals etc) and poorly degradeable wastes (e.g. plastics) is unhelpful if rapid degradation is wanted. Thus a degree of pre-tipping sorting is essential, and this - of course - allows easily removed and economically valuable materials to be recovered.

2 To utilise very much smaller waste cells, which can be filled in at most a few months. If this is not done then the most readily generated landfill gases (from food wastes and organic fines) simply escape to pollute the atmosphere. Since these earlier gases account for half the available gas resources, a large economic loss is also incurred if they escape.

3 To check the moisture contents of whatever sorted wastes are tipped in cells, and modify these, where necessary, to have 60 to 70% of saturation moisture content. If wastes are very dry, biodegradation slows and then stops, and so gas production suffers. Conversely too wet a waste allows finer and soluble materials (the preferred bacterial food stocks) to be washed down the wastes and gas generation falls off. The ideal is to have a high enough moisture content to promote easy movement of bacteria to uniformly attack the waste's degradable content.

4 To ensure that wastes are not overly compacted, since this precludes the easy movement of moisture and bacteria.

5 To install adequate insulation (e.g. 2m of soil as a minimum) above each waste cell to retain the heat generated as wastes rot. An internal cell temperature of 35^{0}C to 45^{0}C is known to give optimum growth conditions for bacterial populations. Lower temperatures give rise to lesser landfill gas generation rates.

6 To monitor conditions in each waste cell to ensure that stable methanogenic conditions are occurring. This is a relatively inexpensive requirement, since a few indicator chemicals adequately typify which degeneration condition is in progress.

The above requirements obviously do impose limitations and costs on landfill operators. Sorting wastes will call for an extra staffing (of perhaps 10 people), a conveyor belt/screen system and a magnetic separator, and so the capital costs will be higher (about £1.5 million for a 2 million cubic metre landfill). Reducing waste cell sizes will obviously reduce the volumes of wastes which can be tipped, as indeed will the limitation on compacting tipped wastes. On a landfill with a 2 x 10^6 cubic metre capacity, these additional costs are likely to amount to about £2 million, in total.

However, the requirements also attract benefits. A traditional landfill can be expected to produce 100 cubic metres of landfill gas (60% methane and 40% carbon dioxide) per tonne of tipped wastes, plus leachates, whilst one of the modified bioreactor type described will produce 250 cubic metres or more of gases per tonne of degradeable wastes, without significant volumes of leachate. Since these landfill gases (in the dried but otherwise untreated state) have calorific values (at 2.1 MJ per cubic metre) very similar to that of town gas, the extra gas resources can be a major economic gain, particularly if these can be piped directly to industrial consumers and sold at some 50% of the costs of mains gas.

A traditional landfill, with a 2 x 10^6 cubic metre storage capacity, can be expected to generate a total income of £22 millions, at 1992 tipping costs, and will have minimal operational costs of perhaps £1.8 millions (for staff and legislative monitoring and compliance) over its 10 year operational life. Thus, even when land costs, and the charges incurred in gaining planning consent, are included, sizeable profits are made by operators. A landfill operating on the revised bioreactor basis will prove to have a very similar balance sheet, if it can be assured that the landfill gases can be sold. The sole economic difference is that pre-landfill investment is necessarily higher, to allow for waste sorting. This should not prove to be a major obstacle, lenders can be assured that future revenue will cover borrowing, since disposal facilities for domestic wastes will remain in demand.

4 NEW CUT HOUSE FARM LANDFILL

Whilst no landfill has, as for as is known, been designed from the start to operate as a bioreactor gas generation system, one at New Cut House Farm (Merseyside) is in the process of obtaining planning permission.

The site (fig.2) permits landfilling over a plan area of some 60ha, and utilises waste cells, each of a capacity to accept only 3 months waste imports. Those cells are 7m high (to promote adequately high internal temperatures), and are only 1/60th of the area of more normal waste cells.

Gas outputs are anticipated to be as shown in Table 1, and the total energy output is calculated as 9,240 million MJ.

Table 1

Predicted gas generation rates - New Cut House Farm

Years after land-filling commences	Gas outputs (millions of cubic metres)
1	0.56
4	29.76
9	49.44
12	62.16
16	72.60
20	69.06
23	33.36
25	0.96

Total gas generation 440 million cubic metres over 29 years.

Total energy equivalence 9,240 million MJ

Economic appraisal indicates that this particular landfill will be almost as profitable as a traditional equivalent, but far less damaging to the local environment.

Separation of received wastes (Fig.3) will give rise to recoverable metals and plastics, though no value has been ascribed to these recoveries in the above economic appraisal, since market conditions for recovered materials are currently in a state of some instability.

The overall economics are such that private sector investors have been able to accept the proposed operation, and no requirement for state funding is necessary.

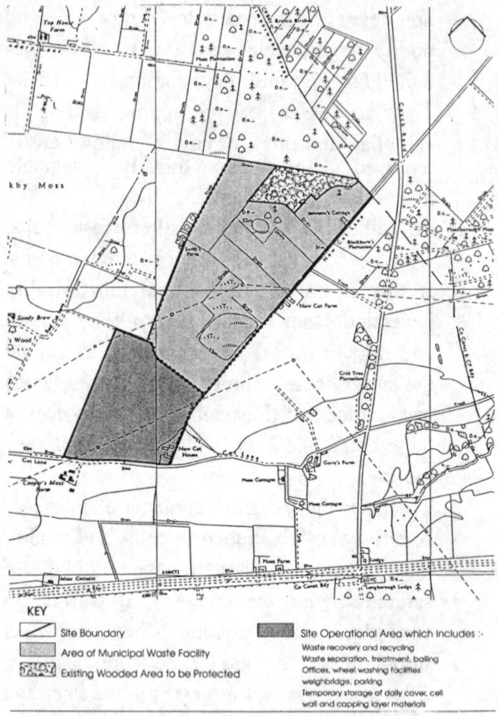

Figure 2 Indicative Site layout

5 CONCLUSION

Varying landfill practices to promote bioreactor gas generation rates offers a cost effective solution to the problems of dealing with domestic refuse, and avoids the usual need for recycling initiatives to obtain scarce public sector investments.

As materials separation is included, a reasonable level of materials recovery can be anticipated, but the main aim is to produce economically useful energy.

This approach avoids the constraint of requiring public start up investment, and is practically achievable. All that is necessary is to bring about changes in traditional landfill operating practices.

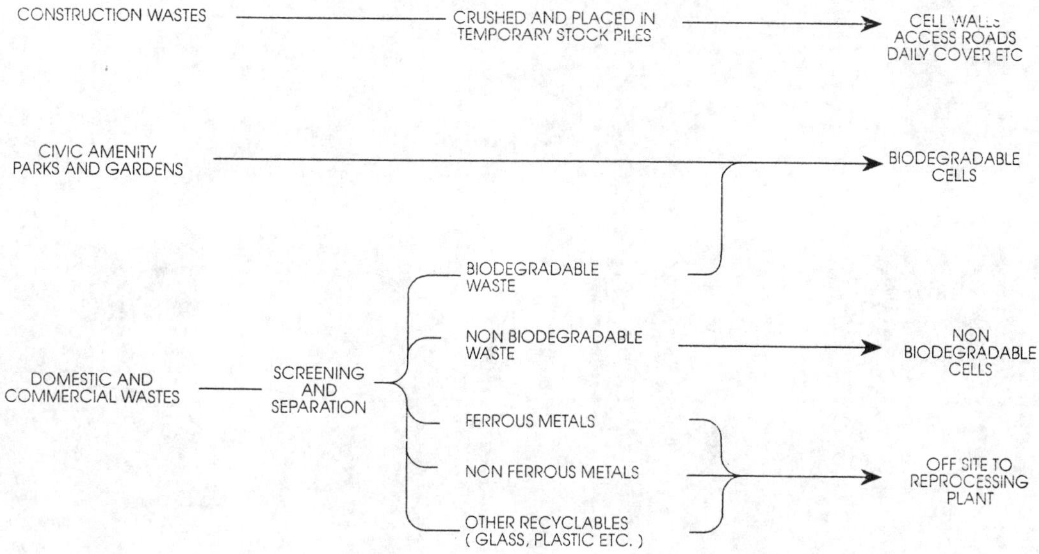

Figure 3 24734 NEW CUT HOUSE FARM KNOWSLEY WASTE SEPARATION FLOW DIAGRAM (Indicative)

REFERENCES

1 Council of Europe 1993. "Report on the management, treatment, recycling and marketing of waste". Strassbourg, France.

2 Environment Business 12th & 26th August 1992. "German waste dumping prompts French import ban". Information for Industry Limited, London.

3 Environment Business 9th September 1992. "DOE sticks to recycling targets despite Halton". Information for Industry Limited, London.

4 Department of the Environment February 1993. "Annual report on resource commitment to environmental protection". HMSO, London.

5 European Commission October 1993. "Proposals of the Belgian Presidency on recycling points system". Brussels, Belgium.

6 Barlaz, M.A., Milke, H.W. and Ham, R.K. 1987. "Gas production parameters in sanitary landfill simulators". Waste Management and Research, Vol.5: pp 27-39.

7 Rees, J.F. and Grainger, J.M. 1982. "Rubbish dump or fermenter?" Process biochemistry 1982, November/December: pp 41-44.

Why Recycle?, Rainbow (ed.) © 1994 Balkema, Rotterdam, ISBN 90 5410 367 1

The opportunity costs of recycling ideology (crawling out of the sink)

Caesar Voûte
Corporation of London, UK

ABSTRACT: This Paper aims to meet the title of this Seminar head on: why are we doing what we are doing under the banner "recycling"? The attempted answer is one from a humanist perspective; a perspective which puts people at the centre. The answer suggests that we will have to rethink "recycling" fairly drastically to remain credible "recyclers". A case is presented, in the form of six propositions for debate, which argues that hard headed, traditional, economic thinking is a necessity, for environmental as well as for social reasons, if recycling is to become more than a symbolic part of society's resources management strategies.

1 THE OPPORTUNITY COSTS

If you buy a new carpet for your living room, the money spent represents an opportunity cost. The new carpet will, presumably, provide some satisfaction and enjoyment, but what other benefits could you have obtained with that money if you had not bought that carpet? We all know this, for we bore each other with sayings like "you can't have your cake and eat it".

If you have 50 pence left in the world, and you haven't eaten for 3 days, buying a cheese sandwich with your money represents, on the other hand, just about zero opportunity cost. It is hard to imagine that spending that 50 pence on a non-food item would have improved your quality of life in any way at all. In this case the saying must be amended to: "you can't have your cake, for you must eat it."

It is not very difficult to find examples of different opportunity cost levels in various waste recycling activities and I will give

examples of different opportunity cost situations.

The first photograph shows one of the most extreme waste recycling situations I know of. It was taken in 1941, in Poland, in the ghetto of Lodz. It shows children excavating part of an old landfill site which happened to be located within the ghetto's boundaries. The children were searching for bits of scrap wood for fuel, pieces of metal, scraps of cloth, anything at all which could be of the smallest use and could be sold for minute amounts of money and thus re-entered the local materials economy [1]. Those children were trying to stay alive. The ghetto's economy valued their lives as equivalent to those little bits of rubbish.

In economic terms, you can say that their time-opportunity cost was effectively zero. Any alternative way of spending their time would reduce their survival chances for the next few days. They had no other option.

This is an example of recycling at zero opportunity cost to both the recyclers and the economy in which it took place. This form of waste recycling has a large and very sad history. It still exists in many parts of the world where people search through waste to try to stay alive.

It is more difficult to capture the opposite situation, of recycling at a very high opportunity cost, in a picture. I have tried to do so by taking a picture of recycling containers sited in the vicinity of extremely high cost housing in London (the location is near the Tate Gallery). The people who live in the flats or apartments in the background will be affluent middle class people who give their time a high value. If they use those recycling facilities, they spend some of that valuable time in doing so, and they get nothing tangible in return.

We can estimate a financial value of their time-opportunity cost of putting a few bottles in the glass bank. In an employed capacity their time will be worth in the order of £20 per hour. They may well value their leisure time even higher, but let us assume the £20 per hour value. It will take about 3 minutes to come down from a flat with, say, 6 bottles and walk to the glass

bank and make the return journey. The time cost of those 3 minutes will be £1. The 6 bottles (assume wine bottles) represent apprimately 3 kg of glass. If you work it out per tonne, this activity seems to have a time-opportunity cost of some £333 per tonne of glass recycled. This is, of course, a very crude calculation, but it shows at least the order of magnitude of this type of opportunity costs. A significant time subsidy is involved.

The containers at that location appear to be well used. The residents at that location clearly believe that recycling is important enough to justify the extra effort and inconvenience of saving the bottles for the glass bank. A more extreme example of this type is the observation that residents sometimes deliver

recyclables to a recycling centre in Camden by taxi. Similar opportunity costs calculations could be carried out.

I have given two extreme examples with the aim of making the basic concept very clear. The third photograph shows an intermediate situation.

Again, the formal economy values the time of this recycling waste sorter at a certain financial level. This time value will be much less than that of professional people of the previous example, whose time value greatly exceeds the economic value of the bottles recycled. In this case the first assumption would be that the waste sorter's time is valued at a rate which makes his employed time cost slightly less than the economic value he adds to the waste as a result of his work.

The waste sorter's employed time could be worth more than his value adding effect on the waste, in which case we have, again, a subsidised recycling operation with its associated opportunity cost.

We should, again, not only consider the waste sorter's externally assigned time cost, but also his time opportunity cost of doing this work from the point of view of his own situation. To what extent does he need to do this work to stay alive? If alternative or better paid work is hard to get, his own time opportunity cost in this work is pretty low. It is not zero, however, if there is still a "fall back" position - e.g. as unemployment benefit, in which case he would have less money, but more time to do things he might like more (or dislike less) than sorting waste. If there is no "fall back" position, his time opportunity cost is virtually zero. Again, analogously with the people of the previous example, the waste sorter could value his non waste sorting time very highly. We know nothing about his background.

It seems to be the case that if we want to retrieve low value recyclables from waste, we need an adequate supply of people who either have very low time opportunity costs in the formal economy or who are willing to incur high time opportunity costs to carry out this work. Such work may, or may not, be subsidised in the formal economy, but there is a real possibility that a substantial "hidden" subsidy is involved, namely in the form of time opportunity costs of those who do the work.

2 THE IDEOLOGY

Waste recycling in a country like Britain is not only sometimes subsidised in the formal economy, but also seems to be subsidised in terms of people's private "sub-economies", and these more hidden subsidies, or opportunity costs, can be substantial. These opportunity costs, where they occur, are intended to purchase environmental benefits. We ought to be puzzled about the absence of a certain question:

What other environmental benefits could have been achieved with the same inputs, or at those opportunity costs?

To put it more colloquially: are we getting the best environmental value for money and human time? Perhaps we are foregoing much greater environmental benefits, for the same input could be used for very different environmental initiatives.

I have never heard this question asked about any recycling scheme or initiative. I hope you will agree that I am discussing an incredibly obvious question, one which ought to be asked all the time if resources are expended for environmental reasons.

If in any system of thought obvious questions are not asked, we have an ideological system. Ideology is that part of peoples' thinking which does not admit alternatives. In this Paper I am assuming that we have a shared ideological assumption that environmental protection is extremely important; the alternative is not up for discussion in this Paper. I am not using the word "ideology" in a pejorative sense. There are benign ideologies as well as evil ones. We do not like our ideological views to be questioned for that makes us insecure. You need ideologies in society anyway, for if you always question everything it is impossible to act.

Ideologies are problematic, however, if they become contradictory. If you set yourself some objective, but one of your ideological beliefs obstructs the achieving of that objective, you have a problem, for it will be impossible to act rationally. It is, therefore, necessary to question ideological beliefs occasionally.

Ideological contradictions are very common. You typically run into such contradictions at the interface between economics and politics, but I will give a simpler example. In the 19th century there were protestant religious sects which aimed to convert society to their view of the true faith, but they also held the view that marriage and sexual relations in general should be avoided. Such sects invariably died out, for they did not bring up their own next generation of true believers from childhood. They had their own built in ideological "self destruct" mechanism.

If the environmental opportunity costs of waste recycling are not considered, we may have a similar problem. The overall aim is the best possible level of environmental protection, but if you hold an ideological view that you must, at all times, recycle waste, you may well be preventing the achieving of much greater environmental benefit because you are excluding the possibility of alternative forms of action using the same resource input.

3 THE OPPORTUNITY COSTS OF RECYCLING IDEOLOGY

The use of high value human time to recycle low value glass bottles in the second example could be a case in point, but there is a better example which shows the magnitude of resource inputs which can be involved. It is contained in recent Government sponsored research on economic policy options to encourage recycling and to meet the government's target of recycling about 25% of the country's household waste [2]. The argument is subtle, and I set it out in steps.

i) The terms of reference required the consultants to consider the measures needed to make the waste disposal process carry its full impact costs.

ii) When waste disposal carries its full impact costs, waste disposal is no longer competing unfairly with recycling in economic terms.

iii) The consultants discovered that meeting the Government's household waste recycling target would incur a net "social cost" of appr. £110 per tonne; that is:

the full avoided waste disposal impact cost (as far as could be expressed in financial terms) has been allowed for. The total net cost of meeting the Government's target would therefore be in the order of £500,000,000 per year. The report summarised the situation as:

"...even when prices reflect the full social cost of final disposal, they may still be below the level required to cover the high costs of collecting materials for recycling to meet recycling targets...In other words, the efficient level of recycling is unlikely to match precisely the required level of recycling.[3]"

iv) If waste disposal costs recycling is still not "efficient" (i.e. cheaper), there is a real risk that subsidised recycling causes worse environmental impacts than waste disposal, for waste recycling processes have their own environmental impacts, which would not be reflected in the financial cost of recycling. In any case, what other, perhaps much greater, environmental benefits can be achieved for the expenditure of £500,000,000 per year?

It seems fairly obvious that the Government's household waste recycling target is an ideological one, for in waste management terms it is insignificant. Only 4% of the country's waste is household waste, so that the target is one of recycling 1% of all waste. This is less than any reasonable margin of error when measuring the country's total waste output.

The findings of the consultants' report have shown that this ideological policy seems to carry an opportunity cost of about £500,000,000 per year. This opportunity cost could well be much higher if we try to measure peoples' private opportunity costs involved in such an enterprise.

4 THE SINK

If you agree that I have discovered a bug in the recycling system, we have to find out where exactly that bug is located in the system. I will try to trace the bug in the basic economics of waste management.

The normal life cycle of an object is:

i) Production: we produce something (e.g. a carpet) which represents an asset. It has a positive value.

ii) Consumption: we use the product, and in doing so it depreciates (the carpet wears).

iii) Disposal: the product no longer serves any useful purpose to the consumer and he wants to get rid of it. It has become a liability.

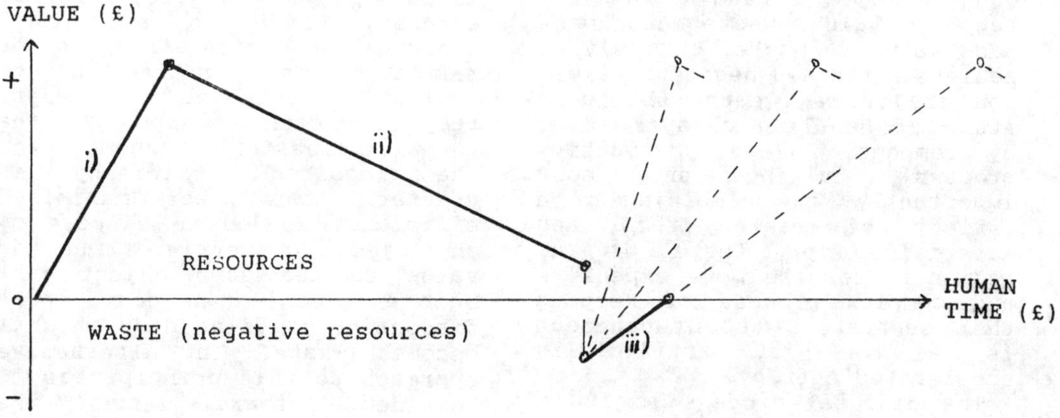

The value change of the object over stages ii) and iii) is not continuous. As long as we still consume that carpet, it still represents some positive value. As soon as we do not want it anymore (stage iii), it is in the way and a nuisance and you have to incur cost or inconvenience to remove it from your home. Its value has suddenly turned negative, without passing through zero. The little graph illustrates the idea, on the basis of the 18th century notion that labour (human time), not nature, is the source of all value [4] and the 20th century realisation of the converse that the (waste generating) consumption of value (goods) is to substitute goods for human time [5,6].

For simplicity, "value" is assumed to be synonymous with "utility". Both value and time are expressible in terms of money, and are plotted with reference to the owner of an article at any stage.

You could use this concept to define waste management as "the management of goods of negative value". It is rational to expend resources on the management of assets (e.g. maintenance, repairs, keeping inventories, etc) as long as those management costs are not higher than the value of the assets. In waste management it is rational to minimise the management costs as much as possible, for the objects you manage already have a negative value. Waste management adds value to waste, but only to achieve a reduced negative value. You try to make each subsequent stage of handling cheaper, e.g. by compaction (a destructive process), bulking, and, most importantly, the combining of a variety of waste items and materials, for there is no rationale for the more expensive and complex process of keeping them separate. Waste management is a form of artificially accelerated entropy.

The disposal process is 100 % successful when the negative value has been increased to zero value, i.e. the waste no longer constitutes a liability to anyone. This is the final stage on the little graph, which is again continuous.

In this situation, waste recycling is an economic tour de force: at some point in the negative value waste management stage you try to add so much value to waste - reversing a degree of entropy already induced by its waste status - that it becomes again an asset. When asset status has been reached, recycling has been completed, and the production - consumption (depreciation) - disposal cycle starts again. This is shown by the dotted lines on the graph which show various degrees of efficiency in the recycling process. Efficiency (the amount of production per unit time) is, throughout, the rate at which human time generates (or depreciates) value; a mathematician would refer to the derivative dV/dT at any point; dV/dT is negative in the consumption stage.

Recycling can only be economically competitive with waste disposal if the recycler(s) can add more value for a given time input than the disposer(s).

If you assume that no mistakes are made in the system (i.e. assuming that people do not accidentally dispose of things which still have a true positive value in a single, formal economy, which is a typical classical economics assumption of market forces operating through rational behaviour), waste recycling cannot happen in the formal, classical economy, for the required efficiency is greater than the maximised efficiency needed to dispose of an object or retrieve and add value to the same object just before it undergoes the discontinuous value drop when it becomes waste. (An alternative approach to this principle is to consider integrals over the various possible systems, defining areas under the curves as "work")

The waste management stage of a product's life cycle is, as a rule, not reflected in its manufacturing or purchasing costs. Waste management is the sink of the system. A system's sink is that part where inevitable losses occur, and what happens to those losses is of no interest to a system's designer or operator. In theory, the entire Universe is the sink of a radio broadcasting system. After consumption (i.e. listening to the sounds) the energy of sound waves as well as of electro-magnetic radiation losses of the system are dissipated into infinite space. You can also think, very concretely, of e.g. the kitchen sink in a household.

In this context, however, we can enrich the metaphor of the kitchen sink by considering the efforts applied by a spider which tries to crawl out of a sink against its smooth and increasingly vertical side. The economics of waste recycling appear to have characteristics, (on my graph) which are similar.

Such recycling could only happen if there are two separate, but slightly interacting economies of a class of rich people, with high time opportunity costs, who substitute goods for time, and one of poor people with very low time opportunity costs, who sacrifice time for goods. The Lodz ghetto, almost completely sealed off from the rest of the economy, is, again, an extreme example in history, but Benjamin Disreali's "Two Nations" and the real waste-scavenging existence of a section of London's population in the 19th century, convey the concept generally [7].

If we want waste recycling to happen in this way we have a choice between the "Two Nations" model and an alternative mechanism which operates outside the formal economy. The reality seems to be a mixture of the two. We have relatively poorly paid (or underpaid? - another time-subsidy) workers in the waste and waste recycling industry, together with the ideologically driven high opportunity cost input by people who could easily, and perhaps justifiably, have spent their time doing something else.

The ideological "bug" makes the system work in a way which "cheats" on the formal economy by hiding the true amounts of value adding (and effort) needed. The price for that form of cheating is the real risk that we do not achieve as much environmental benefit as ought to be achieved for the given resource input.

I will suggest an explanation of why this bug could have evolved. Society's waste management (be it formal, lawful, informal or illegal) is the sink of the system. The point where we begin to reintroduce into the general environment all those things we extracted from the environment and consumed. Just like the kitchen sink, we do not normally want to know too much of what happens in the sink of a process, for it is a negative part of the system. It yields no obvious benefits (we think, simplistically). Once your old carpet is out of your house, and someone else deals with it, or you have poured your washing up water down the kitchen sink, you have little interest in what happens next to these wastes. That is a perfectly normal attitude. We pay, directly or indirectly through taxes, for these invisible services to be carried out properly.

The ideology of recycling looks at the sink as a moral system instead. Retrieving something from the sink makes you feel that you have "saved" something. You have put something right, which was wrong. All this is a complicated way of saying that the recycling ideology is about "feeling good" - a phrase which is used often enough, but nevertheless requires a proper explanation. "saving something from the sink" is a symbolic act which represents "saving the environment". (Can you perceive an analogy with the Victorian preoccupation with the salvation of "fallen women" to "save" society?).

In my book, "saving" something from disaster is the exact opposite of managing a process, which aims to prevent occasions when acts of salvation are needed as much as possible. This applies to economics as much as to the environment.

5 CRAWLING OUT OF THE SINK

I will now offer some propositions for debate to suggest how we can make a transition from ideological recycling to resources management.

Proposition 1.

Members of the public who expend time and effort to participate in recycling activities have a right to know what the environmental net benefit of their efforts is. Without being provided with such knowledge, they cannot know whether their efforts are symbolic only or whether they are worthwhile in a more real sense. Industries which make and distribute goods which are expected to be recycled cannot reasonably expect the public's recycling efforts and goodwill to continue indefinitely without such information.

For example: producers, distributors and retailers of glass bottled beverages cannot reasonably, and realistically, count on indefinite public support for the bottle bank system unless they prove that this form of glass recycling is environmentally justifiable compared with returnable and refillable bottles.

Proposition 2

When we design, make and market an article for consumption, we must make up our mind whether that article is disposable after consumption, or whether it is intended to be refurbished or reprocessed in some way to re-enter the consumption stage. Having made that decision, we must act on it consistently. If we have decided that it is disposable, we must dispose of it responsibly and efficiently, and not subject it to recycling ideology.

For example: is an aluminium drinks can designed for disposal or for recycling? If, for the argument's sake, 40 % of those cans are recovered for recycling, the design aim is 40% successful if recycling was the object, but it is 60% successful if the design aim is one of disposal. The recycling percentage is meaningless until the design aim has been declared.

Proposition 3.

If certain goods are specifically made to be labour saving during their consumption stage, it is irrational to negate those labour savings by requiring the consumers to engage in time consuming activities for a recycling operation after consumption.

Proposition 4.

The profitable recycling of materials from municipal waste is inconsistent with the belief that the market is efficient. One or the other must be the case.

Proposition 5.

Waste recycling corrects market inefficiencies. That is, waste recycling is a corrective measure when resources have erroneously entered the waste management system. An efficient recycler in a rational economy discovers resources of positive value which accidentally found their way into the system of negative valued resources. He retrieves the positive value resources and makes a profit. Since efficient recycling exploits and corrects mistakes, it can only be a temporary and opportunistic activity. As soon as the error is discovered, the positive value materials are no longer disposed of as waste. Waste recycling opportunities diminish as society's management of resources improves. This implies that efficient waste recycling cannot be planned fully, and cannot be an intrinsically desirable activity.

Proposition 6

An entrepreneur adds value to resources to create a product. He typically dissipates that product as widely as possible to consumers in society to be able to derive revenue from as wide a spectrum of society as possible and makes a profit. He dissipates goods and concentrates money. Is it possible for a government to change the economic ground rules in such a way that this entrepreneur can re-assume ownership and retrieve the products after consumption (i.e. reverse the dissipation of goods and re- dissipate money instead), carry out recycling and still make a profit on the full cycle?

If this is in general possible, the problem can be fully solved, for the retrieval of resources takes place before the discontinuous value drop occurs where goods become waste. If this is in general not possible, however, we have to accept that public sector waste disposal is a necessary part of the production-consumption cycle of products and waste recycling has the more limited function of correcting temporary market errors until such errors are rectified.

REFERENCES

1. Adelson A. & Lapides R. 1989. Lodz Ghetto, inside a community under siege. Harmondsworth: Penguin.

2. Environmental Resources Limited. 1992. Economic Instruments and recovery of resources from waste. London: HMSO.

3. Ibid. p.61.

4. Adam Smith. 1776. The Wealth of Nations.

5. Cairncross, Frances. 1991. Costing the Earth. London: The Economist Business Books.

6. Schwartz Cowan R. 1989. More work for mother; The ironies of household technology from the open hearth to the microwave. London: Free Association Books.

7. Henry Mayhew. 1861-2 and (reprint) 1968. London labour and the London poor. New York: Dover.

* This Paper is written in a personal capacity, and does not necessarily reflect the Corporation of London's policies.

Why Recycle?, Rainbow (ed.) © 1994 Balkema, Rotterdam, ISBN 90 5410 367 1

The cost of recycling

Richard Waite
Coopers & Lybrand, London, UK

ABSTRACT: Despite the development of recycling in the UK, there has been relatively little progress in improving information on the cost of recycling. A proposed standard approach to calculating such costs is presented, together with the latest estimates of these costs.

In 1977 there were just 17 bottlebank sites in the UK. In 1992 this number had grown to 7,451 and by the year 2000 the glass industry aims to have 10,000 glass bank sites across the UK. There has been a similarly explosive growth in the number of paper, can and textile banks, and in the last few years we have seen the proliferation of kerbside collection schemes. Given the extent of household recycling which now takes place, and the fact that it all started over 16 years ago, it is perhaps a little surprising that we still cannot answer one of the most basic questions concerning recycling - "what does recycling cost?"

We know that recycling costs more than landfill, but we cannot say how much more. We have estimates and we have ranges of costs, but we cannot yet be definitive.

The key problem is that there is no agreement on how the cost of recycling should be calculated, there is no standard approach. Different schemes report their costs on different bases, so that the comparison of alternative approaches is prevented and the translation of one experience into a different set of circumstances is severely limited.

In order to address this problem, Coopers & Lybrand in conjunction with the European Recycling and Recovery Association (ERRA) have developed a proposed standard method of calculating the cost of recycling. It is hoped that as many schemes as possible will adopt this approach, both in the UK and across Europe, in order to ensure that the true costs of recycling are reported, and different approaches can be compared with confidence. This paper presents an outline of the costing model.

Recycling comprises three stages:

(a) collection - normally by kerbside or bring methods;

(b) processing - that is, bulking-up, storage and transfer or sorting and separation for example at a Materials Recovery Facility (MRF); and

(c) reprocessing - when the form of the material is changed, for example paper pulping, glass melting or organic material composting.

With regard to the cost of reprocessing, this cost is essentially reflected in the market price offered for each material, by the reprocessing industries. Thus, the energy savings associated within the reprocessing of glass or aluminium are reflected in the relatively high market prices offered for these materials. Whilst there will be some market price distortions, as for example in the recent case of paper and plastics, it is reasonable to take the cost of reprocessing as being reflected in the market prices. Any analysis of the cost of recycling need therefore only consider in detail, the costs of collection and of processing.

The costs of collection and processing are directly linked in an inversely proportional relationship: the greater the degree of separation at source, the greater the cost of collection, but the lower the cost of processing, and vice versa. The costs of collection and processing must always therefore be considered together, but for the purposes of measurement and modelling, they are more easily treated separately.

The cost of recycling may therefore be expressed as:

Cost of collection +
Cost of processing
Cost of recycling -
Material sales income
Net cost of recycling

It is important to report both the gross and the net cost of recycling, since the income derived by any scheme from the sale of collected and processes materials, will be dependent on local market conditions. When comparing schemes or when using the cost of one as a basis for planning another, it is therefore important to eliminate such variability by using the gross cost of recycling.

The cost of collection

The costing model which has been developed, divides costs into two categories, which in turn are made up of a number of sub-categories, some examples of which are as follows:

Capital - land/buildings
- vehicles
- containers

Expenses - staff
- transport
- accommodation
- promotion
- other

The capital costs are essentially self-evident: land and building costs associated with vehicle depots, offices etc.; collection vehicle capital costs; and the collection containers such as bins, banks or boxes. Similarly, expense costs comprise: direct and indirect staff costs (wages/salaries, National Insurance, pension etc); transport costs such as vehicle fuel, insurance and maintenance; accommodation costs such as rent and rates, heat, light and power; and the costs of promoting the collection scheme. This latter cost category is often omitted or underestimated, but is significant if a scheme is to be successfully launched and maintained.

In terms of identifying the value of each of these cost elements, what we are interested in is the marginal cost of collection. We have as a base cost, the cost of a refuse collection vehicle visiting every household every week. The marginal cost of collection for recycling is then any collection cost that is additional to this base cost. Thus for example, in a Blue Box scheme, a separate vehicle is used to collect recyclable materials; the full cost of this vehicle is therefore recorded as one of the capital cost elements for recycling. However, in a Green Bin or Bag scheme which uses a divided compactor vehicle to simultaneously, but separately collect refuse and recyclables, only the additional cost of the divided vehicle, compared with a standard vehicle is recorded.

The cost of processing

The cost headings for material processing are similar to those for material collection:

Capital - land/buildings
- equipment

Expenses - staff
- accommodation
- process
- other

For processing, the capital costs may be considerable, for example in the case of a MRF with a high degree of automation. Alternatively staff costs will be high, if a lesser degree of automation is used for material processing. Processing expenses include such items as baling wire, equipment maintenance (a considerable cost in an automated plant) and staff protective clothing etc.

The costing model which has been developed comprises a greater number of cost categories than those illustrated above and also defines exactly what should and what should not, be included in each category.

The cost of recycling

Having calculated the costs of collection and processing separately, they are now combined to give the overall cost of recycling, as follows:

Collection expenses +
Processing expenses
Operating cost

Depreciation: collection +
processing +
Gross cost of recycling

less

Material sales income
Net cost of recycling

The capital costs of collection and processing are added to the operating expenses in the form of capital depreciation to give a gross cost of recycling. The material sales income is then set off against this figure to give the net cost of recycling.

Measuring cost performance

Having calculated the absolute cost of recycling, we now need to relate such costs to the performance of the recycling scheme concerned. This is necessary to evaluate the cost effectiveness of the scheme. Two measures are in common usage:

(a) the cost per tonne; and
(b) the cost per household.

The former is consistent with waste management practice generally, in that weight in the common unit of measure. However, to local authorities and to individual householders, it is also important to report the cost on the basis on which the Council Tax is calculated, that is the cost per household.

It is important at this stage to define what is meant by both "per tonne" and "per household".

At present, the majority of recycling schemes are based on one method of collection and one method of processing. However, as recycling develops, we shall see the emergence of regional processing centres (MRFs for want of any other description), which will be supplied with material from a range of different collection methods. It is also highly probable, that all such collected material will be weighed on delivery to the MRF (primarily as the basis for charging a gate fee), and that all processed material will be weighed on despatch from the MRF (again for payment purposes). It is therefore likely that both the collected and processed tonnages will be known, however, it may not always be possible to link the processed tonnage to a particular collection tonnage.

It is important to calculate the collection cost per tonne, in order to compare different collection approaches, however, it must be recognised that such a measure takes no account of the quality of the material collected. For example, a scheme collecting a large proportion of non-recyclables or contaminated recyclables will have an artificially low collection cost per tonne. Equally, the processing cost per tonne is an important measure to compare for example, different MRF designs, but does not take account of the quality of the input material, and hence the output tonnages.

The following measures should be used:

Collection : $\dfrac{\text{collection cost}}{\text{number of tonnes collected}}$

Processing : $\dfrac{\text{processing cost}}{\text{number of tonnes processed}}$

Recycling : $\dfrac{\text{collection and processing}}{\text{number of tonnes processed}}$

The key measure is the last of these three, as it takes account of the quality of the material collected and processed.

With regard to the cost per household, there are only two measures of importance:

Collection: $\dfrac{\text{collection cost}}{\text{number of households served}}$

Recycling: $\dfrac{\text{collection \& processing costs}}{\text{number of households served}}$

The same comments as with the cost per tonne, regarding the quality of the collected material apply. The number of households served should be the total number of households covered by a particular collection method, not only those participating, since participation is a measure of the success of the scheme and should not be used to artificially lower scheme costs.

Current UK recycling costs

The only data that we currently have in the UK, are reported costs from a number of the leading trials. These figures are not recorded on a consistent basis, some include estimated elements and all are for small scale schemes. These costs are however of value and are presented below. Also included below are the estimates of likely future costs for these schemes, assuming efficiency improvements, the development of new techniques and economies of scale.

Table 1: Summary of UK collection costs

Method	Current (£/tonne)	Projected (£/tonne)
Kerbside	30-120	10-65
Bring	25-30	20-25
Refuse collection	30-50	

Table 2: Summary of projected UK recycling costs

Method	Gross cost (£/tonne)	Income (£/tonne)	Net (£/tonne)
Kerbside	60-95	10-30	30-85
Bring	20-25	10	10-15
Disposal	10-45		

Table 2 illustrates the significant difference in cost between recycling and the alternative of disposal.

Conclusions

In order to generate reliable data on the cost of recycling we need a common method of cost calculation. Such an approach has been developed by Coopers & Lybrand in conjunction will ERRA, and is available to anyone who would like to use it.

It is important to record and report both the gross and the net costs of recycling, to allow valid comparisons between schemes to be made. Such reporting should be on the basis of both the cost per tonne and the cost per household.

Only by adopting such a standardised approach can we make progress in costing recycling, and until we can accurately define such costs, the development of recycling on a large scale will be hindered.

Why Recycle?, Rainbow (ed.) © 1994 Balkema, Rotterdam, ISBN 90 5410 367 1

The catalyst for cost-effective recycling: Effective removal/recycling of putrescibles at source by the Green Cone integrated with alternating weekly collection of dry recyclables and residual waste

J.M.Lees
High Wycombe, UK

ABSTRACT: This paper proposes a system for very high levels of recycling/reduction (c.50%) without increase of marginal costs, often with revenue savings, by incorporating effective reduction at source of all of the putrescible element, currently collected in the domestic waste stream, as an integral part of the waste management process. Widespread introduction of: - reduction at source (the household) of all putrescibles; - separation for alternating weekly kerbside collection of selected dry recyclables such as newspapers/cans and residuals; - high density banks for glass (and maybe textiles); - can recycle/divert over 50% of the waste arising in households with garden or backyard access from disposal, at no on-cost, and often with revenue cost savings. Applied nationally, such an approach would allow c.40% of all U.K. household waste currently disposed of to be directly recycled, with revenue savings. This should be contrasted to the large additional (marginal) costs currently projected for the achievement of the 25% recycling target.

INTRODUCTION

The current scene

Current projections suggest that net marginal costs to achieve the national 25% recycling/waste reduction target will be in the order of £10-12/household or £60-75/tonne. Marginal costs cited by Coopers & Lybrand October 1993 range from £30-110/tonne less £10/tonne recyclables sales income (likely current value).

Bring systems alone can only be expected to contribute a maximum of 10% of the 25% required and in rural areas are not cost effective (Warren Springs October 1993).

Kerbside collection will be required in most areas to achieve the target. The most economic way to collect is, without parallel collection, by alternate weekly collection of:

- recyclables/organics.
- residual waste.

Where authority wide kerbside collection is planned, bring system collection will probably be limited to glass only (better separation/less breakage/simplified sort requirement). At a 50% recovery of a glass content of 10% of domestic waste (5% overall) this leaves 20% of the target to be met by either kerbside collection of dry recyclables and/or processing of currently collected kitchen and garden organic waste at the household or centrally.

Kerbside collection costs for recyclables are relatively unaffected by the likely range of recovery rates and tonnages, which will be low in relation to normal refuse collection weights, unless only collected monthly. (This appears to lower recovery rates significantly.) Fortnightly collection seems to be a good compromise between cost (prohibitive if weekly) and recovery rates.

Central processing of organics collected from the household is a relatively expensive process producing a low-grade end product (compost) of limited market potential and uncertain value, likely to be depressed rapidly if large quantities of compost are produced from kerbside collected organics.

Organics can be processed at the household either by "home composting", limited to garden waste and part only of the kitchen putrescible stream, or by decomposting by use of the Green Cone, which can recycle 100% of the putrescible kitchen waste, and garden waste from smaller gardens. Both of the above need garden/backyards or easy access to garden space.

Many authorities are planning "home composting initiatives", lending or selling composters or the Green Cone at a large subsidy to householders for them to compost or decompost their organic waste, not as an integrated part of waste arrangement but as a separate method of general reduction of waste for disposal.

Where central processing by composting is being tested, authorities such as Dundee C.C., E. Hants D.C., Test Valley D.C., report eventual 80-85% participation in separation of organics, and a high recovery rate. The conclusion is that organics are easy to sort into a designated receptacle, and provided that is all that is required, a significant majority will separate. It is however likely that only a minority will want to compost at home as the majority may not want the end product nor be prepared to employ skill and effort to product it. The Green Cone, which needs no intervention besides adding all food waste, appeals to the majority, unwilling to make anything more than the minimum effort and uninterested in the end product.

WHAT IS PROPOSED

o Effective reduction of all putrescibles at source (the household) by free loan of the Green Cone to the majority of households with suitable gardens/yards within a collection authority by the WCA or even the collection contractor.

o Introduction of alternating weekly collection of dry recyclables and residuals. Recyclables initially collected via a green sack (Cardiff C.C.) or green bin/box (Bath C.C.) supplied by collection authority/contractor .

Selected dry recyclables initially limited to:

- newspapers and magazines (bundled and tied).
- other paper products (bagged).
- ferrous and non-ferrous cans.

o All glass to be collected via appropriate density of bottle banks only.

Newspapers/magazines and other paper products to be baled and sold-on. Little or no value assumed for newspapers/magazines and a negative value for other paper products (low grade reuse or waste-to-energy incineration). Steel and aluminium cans to be separated and compacted if processing is distant.

60% of available paper products/cans assumed separated at the household together with 90% of putrescibles (participation/recovery rising from 70% over say 6 months following introduction of alternating weekly collection).

It may later be expedient to collect all plastic products as well, to be used as a mixed plastics feed-stock. Waste-to-energy incineration is a likely result of DoE directive to industry: July 1993. No sorting into different plastic products. May be sold-on commingled with paper products if a waste-to-energy solution is adopted for both materials.

WHAT WILL RESULT?

1. *collection loads:*

- as projected, Week 1 recyclables load should be 22-24% of the waste stream, 44-48% of the normal weekly load.

- Week 2 residual waste load should be 45-50% of the waste stream, 90-100% of the normal weekly load.

2. *collection costs:*

The replacement of a normal commingled refuse round by a recyclables collection at c.46% of the load can produce large savings. Based on savings of c.40% of the cost of a fortnightly commingled refuse round, there will be scope for collection cost reductions of up to £7/tonne previously collected. At a typical WCA of 50,000 households, 80% with the Green Cone reduction at source and 650 kgs collected waste, savings will be c.£208,000. Even if this is shared with the collection contractor in a negotiated variation to contract until next renewal, there is a likely saving of c.£100,000 in a typical authority from collection economies.

Cardiff C.C. currently collects from c.27,000 households each fortnight per low cost recyclable collection vehicle in a high density urban situation with easy access to depot/MRF (Kirkby - Warren Springs Seminar: October 1993).

3. *residual waste:*

At households using the Green Cone, this will be putrescible free. At non-participating households, it will become somewhat smelly within 14 days, and may produce fly problems etc. in summer. The remedy is simple for those concerned - use the Green Cone. Bins will also be some 50% heavier than before.

4. *dry recyclables:*

By separation of putrescibles in the kitchen, householders will stop contaminating dry recyclables, especially cans, bottles, other paper products etc. in the kitchen bin, with food waste. The fact that all participating householders will already be recycling (putrescibles) and separation at any time of dry recyclables becomes easy, greatly increases recovery rates and sustained participation.

5. *financial outcome:*

There may be no net marginal costs in respect of collection and processing in recycling c.50% of the waste stream or even revenue savings in areas of high disposal costs. (See Exhibit 3).

THE THREE KEYS TO COST-EFFECTIVE SUCCESS

YIMBY - Yes. In my backyard.
FIFI - Fill it. Forget it. (The Green Cone).
KISS - Keep it simple (and) sortable.

Processing/recycling putrescibles "in the backyard" avoids the high marginal costs of collection, and processing (composting or anaerobic digestion) and additional transport, otherwise required to treat organics.

Weight loss in central composting is c.60-70%, multiplying the on-cost per tonne of end product and putrescible feed-stock needs dilution by 50% of Civic Amenity green waste, straw etc. If putrescible waste is collected kerbside, either weekly by divided wheeled-bins/twin-compartment vehicle or fortnightly via a dedicated and expensive organics bin, then it makes it very expensive to attempt later to collect dry recyclables as well - a fraction too many to collect! No fossil fuel expenditure in extra collections, extra transport and processing if processed at home.

It means that the household is actively engaged in the recycling process - at home.

YIMBY: Yes in my backyard.

The Green Cone is effectively a static organics bin. All that is required is to lift the lid and add food waste. As it takes 100% of putrescibles, in contrast to "home composters", there is no need for the housewife to differentiate between cooked and uncooked waste. No regular intervention is needed except emptying every 1-3 years. The Green Cone is unobtrusive compared to both normal sized wheeled-bins or most branded "home composters". Its visible volume at 100 litres is only 40% of that of these bins or composters, and much more likely as a result, to be located near the kitchen-door than a bulky composter. No skill, effort in turning, aeration, watering or product removal is needed.

FIFI: Fill it. Forget it.

The key to high participation and recovery of dry recyclables is not to ask too much of the householder; keep the selection of materials to paper/cans for kerbside collections, bottles for bring systems and remove the chance of contamination of these before segregation (by facilitating 100% of putrescible separation first). If cooked foods still need to be added to the kitchen bin (as with "home composters" but not the Green Cone), this cannot be achieved. Newspapers/magazines (bundled and tied) other paper and cans can be quickly and easily sorted at a simple and low tech MRF. Process costs escalate very quickly once the feed-stock is complicated by glass (3 colours) and different plastic fractions which need sorting.

KISS: Keep it simple and sortable.

ALTERNATING WEEKLY COLLECTION WITHOUT REDUCTION AT SOURCE

What happens if putrescibles are not removed/recycled at source and the collection authority decides to achieve the 25% target by collection of 5% (glass - bring) and 20% kerbside (paper, cans, PET?), using alternating weekly collection?

Wheeled-bins become essential, if waste was previously collected with black sacks, to deal with c.50% increase in weight (75-80% of domestic waste is collected in Week 2: 20-25% in Week 1). In addition, collected weight tends to increase anyway by 20-25% as well when wheeled-bins are introduced, effectively doubling Week 2 loads.

There is a higher risk of contamination/economic degradation of separated dry recyclables.

Contamination of potential recyclables in the kitchen by putrescibles probably ensures they never get separated for collection and processing, reducing potential recovery.

Residual waste is much less pleasant to handle and containers/vehicles need frequent cleansing.

Ultimate disposal is much more expensive (methane/leachate) and all processes of transfer etc. are inherently unpleasant especially as the putrescible element "matures".

Adverse public reaction to holding putrescibles for 2 weeks without an alternative satisfactory method of disposal (the Green Cone) may force authorities, as at Milton Keynes B.C., to revert to normally weekly collections plus a parallel recyclables collection. A very expensive result.

Putrescibles are a major constituent of the waste stream; their removal is by far the biggest single contributor to major weight reductions and they provide the largest single potential disposal saving in the entire process. This opportunity is missed.

WHY CAN THE INTRODUCTION OF REDUCTION BY THE GREEN CONE CATALYSE THE ENTIRE RECYCLING PROCESS AND MAKE SUCH SUBSTANTIAL REVENUE SAVINGS?

We should explore why achievement of the 25% recycling target is always projected to cost so much.

1. Major capital (or revenue - if acquired via lease) expenditure is forecast for purchase of wheeled-bins (2 or even 3 x 240 litre) split or undivided bins, special and expensive "organics" bins, new split-compartment compaction freighters often with high load capacity for Week 2 loads (150% of normal), lower round productivity which will require extra expensive vehicles. Extra organics processing plus transport to plant.

Parallel recyclables collection rounds (blue box/green sack). Weekly collection of residuals.

* Sophisticated high tonnage capacity MRF (Leeds C.C. £1.5m) required to deal with multi-material sorting, as a result of adding glass and/or plastic to the kerbside sort list. Sorting of paper and cans without glass and/or plastics is very simple.

2. In contrast, the proposed system can be operated without wheeled-bins, or in time with a smaller 120 litre bin as standard for Week 2 residual waste and green bins or boxes (non-disposable) in place of disposable green sacks. Existing vehicles can deal with Week 2 loads later, less expensive and more productive recyclables vehicles introduced to replace those designed for general refuse. Substantial marginal collection cost savings are possible rather than increases.

* If the authority collects generally with black sacks, no on-cost is involved in substituting green for black sacks unless the latter are not an authority supply item, and less and smaller green sacks, need to be given out anyway.

* Compaction is not required except where distance to transfer station/MRF is long (rural areas). Cardiff C.C. use low cost flat-bed lorries with wire cages for collecting their recyclables. Absence of PET in the recyclables list facilitates compaction where desirable.

* Disposal savings/credits on full LRMC basis is considerable if applied to 50% of the waste stream, rather than just 25%.

* Higher recyclables recoveries can lead to higher recovered sales income per tonne of waste arising.

SUMMARY

1. For a typical collection authority with LRMC disposal credits/savings at £15/tonne, projected net marginal revenue savings are £2.75/household, compared to c.£10/tonne net marginal costs projected to be necessary to achieve the 25% recycling/reduction target, a difference of £12.75/household. The proposed Green Cone system can achieve 42.7% recycling/reduction at a effective saving of £510,000 (£12.75 x 40,000) per average authority. (See Exhibit 1)

2. This is a typical UK authority situation post-April 1994, and indicates that nationally, some £225m in savings can be made and still achieve c.43% recycling/reduction. As actual revenue savings are projected and most new equipment is obtainable on operating leases funded by those revenue savings, it is practical to consider rapid introduction of the proposed system. When full collection economies can be realised at the expiry of present contracts, a further £2/household savings can be shown, £80,000 per authority, c.£36m nationally, or c.£260m in total with achievement of 170% of the required year 2000 target of 25%.

3. The scale of reduction of disposals can have significant impact in areas such as Hampshire due to the need to replace existing incinerators by 1996 and scarcity of landfill options.

4. Although no plastics are shown to be segregated, it is foreseen that the introduction of waste-to-energy incineration and non-fossil fuel power generation can use a mixed feed-stock containing commingled plastics and "other paper products", both with high calorific values. This would enable a further c.5.5% to be diverted from disposal (less residual ash content) to give an eventual overall level of c.46-47%. Higher MRF costs will be involved even if plastics/ other paper products are allowed to form the "residual element" once newspapers and cans are removed.

5. The proposed system uses an innovative but well tested system (Battelle Institute and 150 U.K. local authorities) for reduction at source and simple low cost/low technology solutions for collection and sorting.

Exhibit 1

SUMMARY MARGINAL COST ANALYSIS

Assumptions:

o 50,000 household authority with 40,000 able to use the Green Cone (garden access).

o 650 kgs/household domestic waste collected at households and via bring system.

o typical Warren Springs assay with:

	%
o Newspapers/PAMs	16.0
o Other paper	17.2
o Glass	9.1
o Steel cans	4.7
o Aluminium cans/foil	0.9
o Putrescibles	27.0
o Balance of waste	25.1
	100.0

o Recovery levels assumed:

o Papers/cans	60%
o Glass	50%
o Textiles/plastics	N/A

o Putrescibles 70% initially rising to 90% with increasing take-up after 6 months with the Green Cones.

o The Green Cone acquired via 7 year operating leases. Prices include starter pack of Accelerator powder, essential only at start-up.

o General refuse collection cost of £35/ tonne or £22.75/household.

o Collection savings at £3.50/tonne (short-term) £7/tonne long-term (new contract).

o Disposal costs assessed at £15/tonne.

Exhibit 2

LOGISTICS

Total collection tonnage:

- 32,500 tonnes (26,000 t = 80%).

80% of households to be supplied with the Green Cone. Assume the average weight collected/putrescible content same as other households.

Putrescible weight in the Green Cone areas:

- 7,020 tonnes (27%).

Non-putrescible weight in the Green Cone areas:

- 18,980 tonnes (73%).

Fraction analysis:	Tonnes	%
o The Green Cone fraction	6,318	24.3

o "Green Sack"/Bin Fraction:
(80% of available households)

	Tonnage (100%)	Recovery (60%)	
Newspapers & Mags	4,160	2,496	
Other paper products	4,498	2,898	
Cans/foil - Steel	1,222	733	
- Aluminium	234	146	
	10,114	6,068	23.3
	=====	====	

o Bring Fraction:	100%	80%	
Glass} from 80% of homes	1,479	1,183	4.6
	====	====	

o Black Sack/Wheeled Bin Fraction:

	Tonnes/p. a.
Non-putrescible weight in the Green Cone areas	18,980
less recyclables collected above	(6,068)
less bring tonnage in the Green Cone areas (80%)	(1,183)

Plus 10% putrescibles not separated in the Green Cone areas	702	
Week 2 weight collected over one year	12,431	47.8
Normal weight collected over one year would be (Week 1 or 2)	13,000	100.0
	=====	

o Weight diverted from disposal:

	Tonnes	% overall
- in the Green Cone areas	13,674	52.6
- overall 50,000 households	13,870	42.7

Exhibit 3

MARGINAL COST ANALYSIS

	Tonnes	Possible Recovery of Recyclables 60% £
1. Revenue Costs:(1)		
Operating leases for the Green Cone @ £5.00/year x 40,000 (including one starter pack of Accelerator)		200,000
Process/depot costs recyclables(2) @ £11/tonne	6,068	66,750
Glass "bring" costs(3) @ £37/tonne	1,479	54,723
Supply cost - green sacks:(4)		(28,000)
Total Revenue Costs		c. 293,500
		======
2. Revenue Recoveries:		
Collection cost recovery(5)		91,000
Disposal saving LRMC savings at £15/tonne(6)		

	Recovery(7)		
- putrescibles	90%	6,318	
- newspapers }		2,436	
- other paper products}	60%	2,698	
- cans: - Steel }		733	
- Al (+ foil)}		146	
- glass - via bring	50%	1,479	
		13,810	208,000
		=====	======

Sales value recovery (8)

	£/Tonne	
- newspapers	-	-
- cans:- Steel	20	10,745
- Aluminium	500	73,000
- glass	23.50	34,756
- other paper products	(5)	(13,490)
		105,000
Total Revenue Recovery		404,000
3. Net Revenue (Costs)Savings		110,500
		======

SAVING PER HOUSEHOLD c. £2.75

Exhibit 4

Notes:

1. Lease based on a price/Cone inclusive of Caddy/starter pack of Accelerator of £28.60 at £175/1,000 p.a. lease charge over 7 years.
2. Process costs based on simple stand-alone facility to handle 6,068 tonnes (collected kerbside) plus 1,479 tonnes (via banks) or c.7,550 tonnes. Glass already colour sorted. Most newspapers/magazines bundled/tied. Cans in separate bags. Requires magnetic separation and compaction. All paper products baled for dispatch.
3. Bottle banks costs assumed ownership/ operation by WCA or under contract to collection contractor.
4. Bag supply cost based on:
 - existing black sack collection: 80/year supplied free. 40 black sacks replaced by 26 smaller green sacks, or by a "green bin" (similar to Bath C.C.) at equivalent annual revenue cost. Saving: 14 x 5p x 40,000 = £28,000.
5. Collection cost savings:
 Short-term: based on 10% overall cost reduction @ £3.50/tonne collected currently.
 Long-term: based on 20% overall cost reduction at @ £7/tonne collected currently.
 Short-run savings:£91,000 }26,000/tonnes
 Long-run savings:£182,000}

We have assumed that 50% of savings are retained by the contractor until contract renewal.

6. Sensitivity analysis on percentage putrescibles separated indicates an annual loss of disposal credits/savings equivalent to £21,000/year if only 70% of putrescibles are separated.

7. Recycling/reduction level in the Green Cone area = 13,674/26,000 or 52.6%. Overall reduction 42.7%. Assumes only glass banks in 20% non-Green Cone area.

8. Prices ex. MRF.

Why Recycle?, Rainbow (ed.) © 1994 Balkema, Rotterdam, ISBN 90 5410 367 1

Waste minimisation and waste auditing by companies

P.C.Coggins
Centre for Waste Management, University of Luton, UK

ABSTRACT : The paper reviews the role of waste minimisation by industry as part of overall government policies towards waste management and resource conservation. Policies towards resource inputs, process modification and changing product outputs are linked to broader aspects of environmental and quality management and the role of education and training.

Since 1988, considerable attention has been paid to issues of waste management and waste recycling in the U.K. . The emphasis so far has been on household waste recycling, and the role of Local Authorities in preparing Recycling Plans and moving towards meeting government recycling targets by the year 2000. Recently, however, the focus has begun to shift towards waste management by commerce and industry, and this Paper reviews the present situation with specific reference to company waste management strategies, and in particular, towards waste minimisation.

Waste Management Paper No. 28, published to guide Local Authorities preparing recycling plans, recognised the importance of a broad hierarchy of preferred waste management options, outlined below. The E.C. waste hierarchy is very similar, and the Agenda 21 Document which emerged from the Earth Summit in Rio in 1992 stressed this concept.

waste minimisation or source reduction
re-use
materials recycling
incineration with energy recovery
incineration
landfill

In the context of household waste recycling, the authors of Waste Management Paper No. 28 emphasised that it was clearly better to minimise waste and/or re-use products, although these activities were largely controlled by industry. A more extreme view is that the first option in the hierarchy should be waste prevention or avoidance before waste minimisation or source reduction . In this context, the present Paper is more concerned with this aspect of industrial or commercial waste management, using a waste auditing methodology and monitoring waste reduction/minimisation from input materials through manufacturing to product design and marketing.

In the Materials Reclamation Weekly of 5th December 1992. the U.K. Environment Minister is quoted as saying that ' although the government advised industry on how to reduce and minimise waste, it had no plans at present to set specific targets'. If such targets were to be implemented, there will be a knock-on effect in terms of Local Authorities preparing recycling plans and aiming to meet government targets of "... recycling half our recyclable household waste by the end of the century. That is around 25% of all household waste" (This Common Inheritance, 1990).

This position changed in July 1993 when the Department of the Environment invited the packaging industry to prepare plans for recovering and/or recycling packaging materials. The target of between 50% and 75% was mentioned, and the industry was asked to submit detailed proposals by the end of October 1993.

In June 1993 the Department of Trade and Industry launched a waste reduction project (Project Catalyst) in the Mersey Basin involving 15 companies, within the context of suggesting that solid waste and energy consumption could be reduced by between 10 and 30%, water consumption by 20-50% and effluent charges by between 60-80% (Guardian, 14th June 1993). This led on from the waste minimisation/pollution control project based on the Aire and Calder rivers in Yorkshire, initiated by the Centre for Exploitation of Science and Technology in 1992, and involved eleven companies (Wastes Management May 1993, p.33).

Waste Collection Authorities in the U. K. were given a statutory duty to prepare Recycling Plans, and these were to be submitted to the Secretary of State for the Environment by 1st August 1992. These Plans were to itemise waste arisings for household and commercial sources, and were to outline proposals for establishing recycling progra-mmes for household waste. Nothing further was said about commercial waste, although any programme of waste minimisation would reduce baseline waste arisings and also reduce the quantity of potential materials likely to be available for recycling.

In addition, any waste minimisation programme by industry and commerce can have important effects on broader aspects of waste management and resource conservation. In preparing this Paper, reference is made to the emerging range of reports and texts now available on this subject (BiE 1991,1992, Charter 1992, Crittenden and Kolaz-kowski 1992, Department of Trade and Industry 1992, Hopfenbach 1992). In contrast, various reports on recycling published during the last two years tend to concentrate per se on post-consumer diversion and recovery of recyclables, with fewer referring specifically to commercial recycling (CBI 1991).

Alongside these reports on waste minimisation and environmental management, there has been considerable activity recently in the field of 'education for environmental responsibility' (BTEC 1993, Toyne 1993). At a global level, the Earth Summit in Rio in 1992 led to a wider debate on issues of sustainability (LGMB 1993), and these will be referred to in the Paper.

In addition to the emphasis here on volume and/or weight reductions in solid waste, additional environmental benefits may result from minimising waste discharges to air and water, although these are not discussed in detail in this Paper. Effective waste minimisation policies, however, must include necessary controls over waste emissions to land, air and water.

From a company perspective, such approaches to source reduction are associated with concepts of environmental image and the promotion of positive employee attitudes towards environmental issues. Reference is made increasingly to the role of 'stakeholders' and their perception/role in influencing company policies. Stakeholders include employees, shareholders, banks and financial institutions, and other companies linked through material flows or ownership characteristics. In addition, employees may be influenced by their children - who are likely to be faced with environmental issues in the school curriculum. In reality, the arguments may usually be more concerned with cost reduction and/or profit growth, together with a balancing of long term gains against short term expediency.

However, data on quantitative solid waste arisings form industrial and commercial sources is extremely limited. Aggregative data may be available at Local Authority level, but may not be based on weight data. Recent research has attempted to use types of waste handling containers, but the data is unreliable. Attempts to use SIC categories and relate waste arisings to employee numbers have been equally difficult to quantify and replicate (Aspinwall and Company 1993).

One approach to this waste management problem is to consider the internal flow of resources, materials and waste through the production chain. Waste minimisation by industry and commerce, including source reduction, may involve a number of different operational activities, and it will be important to quantify any weight changes influencing subsequent baseline data for Local Authorities. Three subdivisions related to the conventional manufacturing process are outlined in Table 1, although this threefold distinction could equally be applied to commercial or service organisations. In addition to these examples, overall good housekeeping and correct materials handling procedures can be very important in managing and minimising waste in the workplace. Perhaps the simplest amendment to any corporate management policy is to recognise this flow chart, attempt to quantify the various inputs through careful monitoring, identifying emissions and then use such a systematic approach to reduce and/or prevent waste. The emphasis in the table is on solid waste, but ancillary emissions to air and water can be easily incorporated.

Input Changes : Resource Management.

From a supply chain analysis viewpoint, companies will increasingly need to examine the sources and nature of their raw material and partly-manufactured inputs. This will probably involve an evaluation of supply chain companies, and whether they fit within a context of contract compliance regarding environmental/waste legislation. Subsidiary companies may be easier to monitor and control, but companies operating within a larger conglomerate may be subject to additional pressures. With added pressures posed by environmental liability, companies will need to verify the environmental credentials of their suppliers.

Ideally, companies will move towards more efficient use of raw materials, using higher quality inputs where available, and evaluating input substitution with respect to price and quality. Constraints may be placed on supplier companies with respect to contract compliance relative to legislative requirements.

Attention will also need to be directed towards the

Table 1: Source Reduction or Waste Minimisation Strategies

Input Changes : "Resource Management"
- using less materials for the same output,
- using higher quality raw materials, resulting in less waste,
- using different materials, ie. input substitution which may be related to changing prices and/or process modifications,
- using fewer and/or more homogeneous raw materials,
- using less hazardous or toxic materials as inputs,
- using input materials which are more easily recyclable,
- using minimum percentages of recycled materials, but distinguishing between in-house and post- consumer sources.

Process Changes : "Design & Manufacturing Management"
- 'clean processes and clean technologies',
- cost and compatibility of any new equipment,
- producing less waste during the production process,
- good housekeeping :materials handling and logistics,
  - equipment maintenance,
  - prevention of emissions,
  - waste segregation and storage,
- more in-house recycling of materials,
- processing and/or sale of wastes as by-products.

Product changes : "Product Management & Marketing"
- lightweighting,
- product substitution, e.g. plastic for glass, steel for aluminium in drinks cans,
- single material construction as opposed to composite products,
- larger containers, economy-size items, bulk purchases,
- less sales packaging, especially double packaging,
- ease of recycling :extraction by recyclable category,
  - ease of disassembly,
- re-usable, returnable products,
- better quality, longer life and more reliable products.

range of raw material inputs , and their potentially hazardous nature. Using fewer raw materials and manufacturing more homogeneous products are likely to be more resource efficient, whilst reducing hazardous inputs or raw materials with toxic by-products are more beneficial in terms of overall environmental impact.

Policies concerning raw material inputs may be influenced by international trade opportunities, especially where a country may have have limited supplies of high quality domestic raw materials. In the U.K., steel, aluminium and paper industries depend on imports, whilst North Sea oil provides at least an input into petrochemical feedstock, and the glass industry has extensive domestic supplies of suitable sand.

Using more recycled materials, or materials which can be easily recycled, instead of raw material inputs per se must be a key policy objective in 'closing the loop' for recycling within the context of waste management as a whole, and facilitating the concept of "sustainable recycling". For many industries there is a long tradition of using in-house materials to feed back into the production process, although it is sometimes difficult to identify the exact proportion compared to post-consumer inputs. In-house waste is likely to be cleaner, often containing less contamination, and is usually more homogeneous than materials available through post-consumer recycling programmes.

For many recyclables, savings in energy are often quoted as a major factor in promoting their use. Whilst this may be true in comparison with the use of virgin raw materials, the implementation of a 'Carbon Tax' may accentuate the energy advantages of using such inputs collected through post-consumer recycling programmes - although the energy costs in collecting and processing such recyclables will also be influenced by such a tax.

Problems, however, remain in terms of how to distinguish between post-consumer materials and in-house recycling. A broader issue is the whole definition of 'waste', in terms of its origin and potential for re-use - companies may need to recognise and appreciate more explicitly that ' waste represents resources which are in the wrong place at the wrong time'. Increasingly, companies are referring to the use of 'recycled materials' as marketing tools in promoting products, in addition to broader use of terms such as 'environmentally friendly'.

In addition, companies will need to address issues of 'transit packaging', and whether this should be taken back by the supplier companies, or stored separately for recovery and recycling. Pressure from Europe (European Commission 1992) is already forcing companies to assess both the nature of such packaging and instigate recovery and recycling programmes. Some companies are already moving away from boxes or pallet-based systems to on-site silos and container shipments involving no transit packaging (DSM Magazine 1993).

Process Changes :
Design & Manufacturing Management.

This area involves two distinct aspects of waste minimisation and waste management. In the first instance, the development of new cleaner technologies and new techniques may result in the use of less raw materials and/or more efficient use of such raw materials and/or the use of recycled materials. Any gains from such raw material substitution must be weighed against the cost of installing new equipment, and any production inconvenience caused through the installation of new machinery and any changes in production management. For several years companies have been encouraged to use equipment and techniques which will minimise environmental impact - BATNEEC (Best Available Technology/Technique Not Entailing Extra Cost) - and it is likely that major improvements will continue to occur in this area of waste minimisation, although implementation of such new techniques will also involve time constraints.

Similarly new areas of industrial innovation may occur, and indeed be encouraged, in the field of clean technology, and using a greater proportion of recycled materials instead of virgin raw materials. Such industries may represent examples of 'post-industrial' commercial development, and there is evidence from North America of Local Authorities actively encouraging the establishment of such companies.

Good production management housekeeping may lead to overall waste reduction and more efficient waste management without the use of new equipment, and in some instances may even prevent waste in the first place. This includes regular equipment maintenance programmes, and more effective materials handling and logistics. The concept of 'just-in-time' management can equally be applied to the subject of collecting/handling/processing raw materials and any associated waste products, especially where inputs and on-site storage cost money and inadequate storage may involve health and safety issues.

Whilst solid waste may have been traditionally collected on-site and prepared for disposal off-site, there is a greater incentive for segregation and storage into identifiable waste streams.This will still have to be disposed of via an accredited waste carrier (now a statutory requirement under the U.K. Environmental Protection Act) , whilst some of the waste may have a value - in some cases being collected at zero cost to the firm.

New waste segregation policies should encompass office and canteen areas as well as the production unit. In addition, the company may decide to handle all, or some, transit packaging itself, and perhaps in conjunction with a waste disposal company, install a baler within a broader solid waste management policy.

The Environmental Protection Act 1990 introduced the role of registered carriers of (solid) waste, and any such waste arisings must be suitably certificated and consigned to registered carriers for proper disposal. The courts are already enforcing this legislation with fines, and even jail sentences, for non-compliance. Similarly permits for emission to air and water are increasingly being made more stringent and enforceable through the principle of Integrated Pollution Control, and implemented by HMIP and the National Rivers Authority.

Whilst companies have traditionally paid private contractors or the Local Authority to dispose of of its waste, separate collection of recyclables may be possible at lower prices.

An alternative approach for some of the 'waste' materials may be to participate in a waste exchange scheme. Such schemes were tried in the U.K. in the 1970's, but problems of cost, confidentiality and bureaucracy resulted in many of them disappearing (Tron 1990). A number have re-emerged during the early 1990's, in line with renewed interest in the environment, and often associated with school/children's 'scrap store' initiatives.

Product Changes :
Product Management & Marketing.

Consumer products often reflect resource management and waste-related issues. Lightweighting is an obvious example, along with product substitution for drinks containers, although in some cases the resulting products may be less robust and have shorter lifespans. The one-pint glass milk bottle in the U.K. weighed 538 g in 1940, 340 g in 1960 and only 245 g in 1990. Less extreme, but equally significant, have been the lightweighting of aluminium and steel cans and PET bottles.

Specific marketing policies may reinforce or contradict such issues. Larger containers are more resource efficient, and multi-packs may include unnecessary additional packaging materials at the point-of-sale. Examples are the multi-packs of beverage cans with a ' Hi- Cone plastic necklace' - although recent initiatives have emerged to recover and recycle these items.

Purchasing policies by the general public may increasingly influence the use of such post-consumer materials, although careful auditing will be necessary. The use of the Eco-labelling scheme may assist purchasers in distinguishing product claims based on post-consumer waste or in-house waste, but evidence already suggests that some consumers may seek to purchase items which are recyclable, or made out of recyclable materials.

Consumers may increasingly seek products which can be re-used, refilled or recycled, and thus deliberately avoid items which would otherwise only enter a waste disposal pathway. This will more education and publicity for the general public, especially in terms of container composition via easy-to-understand labelling. For consumer durables, the emphasis may increasingly be on better quality, longer lasting and more durable products.

The concepts of Cradle-to-Grave Analysis or Life-Cycle-Analysis are increasingly being used to evaluate individual consumer products, and tend to be associated with such concepts as green companies and/or green marketing. At issue is the integrity of the product in terms of being environmental friendly or environmental friendlier or suffixed by some other environmental description involving benign processing and/or benign inputs and/or benign waste outputs. Progress towards Eco-labelling may, or may not, clarify the various conflicting claims made by manufacturers in this area.

Waste reduction and waste minimisation strategies also need to be placed within a broader context of environmental and waste management legislation. The Environmental Protection Act 1990 laid down various statutory duties for waste producers regarding compliance measures, Duty of Care and licensing of waste carriers. In addition, the publication of BS 7750 on Environmental Management clearly outlines the role of environmental management alongside quality management, and this is supplemented by E.C. moves towards eco-management and relevant standards. The impact of any large scale development on the environment external to the firm is also covered with respect to European and U.K. legislation relating to environmental impact statements.

With reference to BS7750 and waste management it is important to recognise important key stages in developing an environmental strategy. An initial overview of operations and a waste audit will provide baseline data, and this is probably best collected systematically in terms of data on waste arisings and waste containment together with details of existing waste handling, waste transfer and waste disposal arrangements. From this base, existing policies can be evaluated and linked with objectives and targets relating to compliance constraints on the one hand, and also environmental and sustainable issues at a more general level. Having established such a policy, regular review can monitor progress and allow policy modification and fine tuning to take place.

In formulating and implementing such a corporate policy towards waste management and waste minimisation as part of broader environmental management, top management must be involved and there must be a commitment from senior executives. Recent environmental legislation involves the designation of named senior management individuals to be responsible for specific actions carried out by the company. More companies are now making such corporate environmental policies available to a wider audience, and also including key principles within company Mission Statements. As such, environmental management is becoming a component of Total Quality Management.

Implications for working practices mean that all employees must be clearly aware of their individual responsibilities, and this will require a programme of structured education and training. This can be undertaken either in-house using informal methods, or using a more structured approach involving external agencies - indeed, there is an increasing need to ensure independent verification.

Complementing such education and training, many in-house company recycling initiatives to collect beverage cans and office paper often come from employees, usually linked with local or national charities. These are often very important in reinforcing concepts of employee responsibility and awareness of environmental and waste issues

The context of waste and environmental management is thus broadened into the subject of environmental responsibly, raising issues of education and training and the links between industry and education establishments. The Toyne Report (Toyne 1993) dealt with the implications for further and higher education, and the importance of partnerships between industry and education in promoting and providing such environmental responsibility. Reference has already been made to the need to involve all employees in implementing any waste minimisation strategy, and this can be done a number of ways. There will be a need for a range of induction courses for new staff, awareness programmes to accompany changes in corporate policy, and specific training courses with respect to specific compliance measures or the implementation of new technologies and/or new operating practices. In addition to industry and education working together, examples of good practice can be identified and disseminated and waste audit manuals devised for different industries.

CONCLUSIONS

Solid waste minimisation must be placed within a narrow context of waste auditing within companies, and within a broader context of environmental auditing. Wherever possible, prevention is better than cure, and waste reduction desirable through clean technology. Whilst public pressures for action, and legislative responses imposing compliance limits have increased over the last few years, the methodologies and management systems for implementing such procedures are still in their infancy. With problems of economic recession, falling sales and difficulties of arranging finance, some companies have chosen to ignore such pressures.

Data management, ranging form agreement on definitions through measurement and evaluation are still being hotly debated - especially where any allocation of costs and benefits are involved. The balance between economic, energy and environmental costs are still unclear, let alone issues of political accountability and the contentious question of defining and quantifying sustainability.

Whilst discussion has centred in some quarters on validating the external consultancies who can conduct environmental auditing, rather less attention has been paid to issues of internal company changes necessary to manage, monitor and implement any waste minimisation strategies . Waste auditing at company level may be much simpler to undertake, and will not require the services of external consultants. Policies must be implemented from the top, and there must be full awareness at all levels within the company as to the importance of waste reduction and waste management. BS 7750 is a major step in the right direction, along with the European proposals for eco-management, but the broad issues of environmental management will need to be tightened with respect to specific aspects such as solid waste and other environmentally adverse impacts.

Integrating these strands are the education and training needs to foster environmental responsibility at all levels in industry and commerce, and to contribute towards sustainable development. Whilst large companies may be able to implement suitable in-house training programmes, the greatest pressure may come from small and medium sized establishments who do not have the resources nor the staff to commit to such training. External help and the provision of self-help packages may help here.

REFERENCES

Aspinwall and Company Ltd. (1993). General Industrial Waste Programme : Development of a Sampling Methodology. ETSU B 1357,

British Institute of Management (BIM) (1991). Managing The Environment. Results of 1991 Survey.

BTEC (1993). The BTEC Environmental Initiative.

Business in the Environment (BiE) (1992). Your Business and the Environment : A D-I-Y Review for Companies.

Business in the Environment (BiE) (1992). A measure of commitment : guidelines for measuring environmental performance.Charter M. (Ed). (1992). Greener Marketing : A responsible approach to business. Greenleaf Publishing.

Coggins P. C. (1993). The Important Option of Waste Minimisation. Waste Management Proceedings, April, pp. 9-12.

Coggins P. C. , Evans G. H. and Cooper A. D. (1992). Obstacles to Successful Composting. Wastes management, Vol. LXXXII, No. 2, pp. 65-71.

Confederation of British Industry (CBI) (1989). Managing Waste : Guidelines for business.

Confederation of British Industry (CBI) (1991). Towards a recycling culture.

Confederation of British Industry (CBI) (1993). Reducing the burden of waste : Guidelines for business.

Crittenden B. D. and Kolazkowski S. T. (1992). Waste Minimisation Guide. Report for the Institution of Chemical Engineers. With the support of the Department of the Environment.

Croner's Environmental Management Special Report (1993). Issue 1, Environmental Training.

Department of the Environment (1991). Recycling : A Memorandum Providing Guidance to Local Authorities on Recycling. Waste Management Paper No. 28. HMSO.

Department of Trade and Industry (1992) Cutting Your Loses : A further guide to waste minimisation for business. H.M.S.O.

DSM Magazine (1993). DSM silos instead of packaging waste. May , number 49, pp.4-5.

Environmental Resources limited (1991). Deposit/Refund Systems For Beverage Containers And Batteries. Report for Department of Trade and Industry and Department of the Environment. HMSO.

European Commission (1992). Proposal for a Council Directive on Packaging and Packaging Waste.92/C 263/01.

Hopfenbeck W. (1992). The Green Management

Revolution : Lessons in Environmental Excellence. Prentice Hall.
Local Government Management Board (LGMB) (1993). Agenda 21 : A Guide For Local Authorities In The U.K. Local Agenda 21, Earth Summit : Rio, Supplement No. 2.
Netherlands Ministry of Housing, Physical Planning and Environment (1988). Memorandum on the Prevention and Recycling of Waste. Plus updated Present Situation Report of 1989.
Rathje W. and Murphy C. (1992). Rubbish : The Archaeology of Garbage. Harper Collins.
RSNC The Wildlife Trusts Partnership (1992). Guide To Community Composting.
SERPLAN (1991). Waste - Its Reduction, Re-Use and Disposal : A Regional Perspective (Revised Guidance). A discussion document by the Waste Disposal Working Group.
SERPLAN (1992). Waste - Its Reduction. Re-Use and Disposal : Regional Waste Planning Guidelines.
SWAP (1993). Don't Waste Your Rubbish : Your Guide to the Three R's - Reduce, Re-Use, Recycle. For Leeds Environment Department.
Toyne P. (1993). Environmental Responsibility : An agenda for further and higher education. Report of a Committee on Environmental Education in Further and Higher Education, appointed by the Department for Education and the Welsh Office, H.M.S.O.
Tron A. R. (1990). The Role of "Waste Exchanges" in Waste Utilisation. Warren Spring Laboratory. LR 742 (MR).
U.S. Environmental Protection Agency (1990). Waste Minimization. Second Edition.
U. S. Environmental Protection Agency (1992a). RCRA Waste Minimization Action Plan.
U.S. Environmental Protection Agency (1992b). The Consumer's Handbook for Reducing Solid Waste.
U. S. Environmental Protection Agency (1992c). Green Advertising Claims.
Warmer Information Sheet (1993). Glass Recycling.

Why Recycle?, Rainbow (ed.) © 1994 Balkema, Rotterdam, ISBN 90 5410 367 1

Sustainable waste management – Possibility or pipe-dream?

P.J. Mobbs
Mobbs' Environmental Research, Banbury, Oxfordshire, UK

ABSTRACT: The current promotion of recycling by Government and local authorities is not tackling the central problems of waste in the UK; it is un-sustainable in its present form, and does not solve the current waste management crisis. Current recycling targets are unrealistic when compared to the annual UK waste arisings, and the lack of a national waste management and recycling strategy prevents any organised solution.

1. THE SUSTAINABLE MANAGEMENT OF RESOURCES

The industrial world recently signed up to a programme of radical change in the way global resources are managed and used, through the UN Conference on Environment and Development, and the final conference document (UNCED 1992), Agenda 21 (the Agenda for the 21st Century). This requires signatory states to move towards 'sustainable' economic and social systems during the first part of the next century. Given this requirement - and the ever present prospect of increasingly scarce natural resources, can current practices - and assumptions - about waste management and recycling be carried forward into the next century without serious challenge?

The problem with 'sustainability' is that there is not one concrete definition of the term. One of the best analyses of sustainability issues and the development planning system has been produced by the Town and Country Planning Association (TCPA 1992):

"Sustainable development is a vague concept that, at once, offers a comprehensive, consensual and conservative approach able to weld together quite disparate and conflicting interests in environment and development. But, because it is vague and its implications poorly understood, in practice it offers few clear solutions. Anyone can sign up for sustainable development so long as it requires no specific commitment to do anything that will threaten their material interests."

The meaning of sustainable development is normally explained by the oft-quoted definition in the Brundtland Report (UN 1987):

"Sustainable development is development that meets the needs of the present without compromising the ability of future generations to meet their own needs."

In terms of waste management, what the industry deals with is the 'sterilisation' of resources, by burying them, burning them, or dumping them at sea, preventing their use by future generations. Although these resources can be remade, this requires large amounts of energy, and the consumption of more resources. What the waste management industry must achieve, in association with society and industry at large, is the closing of the 'loop' - turning the currently linear use of resources into a cyclical one. This could be achieved through:

1. The minimisation of the amounts of waste/resources being lost to landfill, incineration or other forms of dumping.
2. The highest levels of recycling, where it can be shown on a life cycle analysis that this saves resources and energy.
3. Where wastes cannot be recycled or reused, urge substitution for materials which can.

Preventing waste amounts to the conservation of resources. It involves changes in production and design, with greater emphasis placed on quality, durability and utility. There must be a shift from a linear to a circular production process, instead of a succession of stages of production, each producing wastes that have to be managed in some way. This can be achieved at each stage by a combination of re-use, alternative use or recycling of materials otherwise unused during production and, wherever possible, by the use of waste products for some other useful purpose.

The consumption end of the cycle is the most important - and the hardest in practice - link to close. Products must be either returned and reused, put to secondary use, or recycled back to raw materials. This entails some sort of collection and return system, such as that now operated in Germany for some types of waste.

2. WASTE MANAGEMENT OPTIONS

Currently, waste materials have one of four fates:

1. 'In house' recycling as part of a production process.
2. Recycling of waste products after their useful life has finished.
3. Disposal to landfill.
4. Disposal to incineration (with possible energy recovery).

All these routes have their draw-backs which need to be considered when considering the advisability of which route should be selected for which type of waste. The merits of these various options are outlined by the DoE in Waste Management Paper No.1 (DoE 1992a)

The estimated annual waste arisings in the UK are

Table 1. UK annual waste arisings

Waste type	Arising, Mte/yr	%	Reclaim, Mte/yr	%	% of all arisings.
Mine, quarry, slag and ash					
Colliery & slate	51	13.4	0.5	1.0	0.13
China clay	27	7.1	1.5	5.6	0.39
Fuel ash	13	3.4	6.6	50.8	1.73
Furnace slag	4	1.0	4.0	100	1.05
Other quarrying	9	2.4	--	--	--
Total:	104	27.3	13.1	12.6	3.43
Agricultural	80	21.0	--	--	--
Industrial					
General wastes	50	13.1	--	--	--
Special wastes	2.5	0.7	--	--	--
Total:	52.5	13.8	--	--	--
Dredged materials	43	11.3	0.0	0.0	0.0
Sewage sludge	35	9.2	0.0	0.0	0.0
Building wastes					
Building waste	24	6.3	11.0	45.8	2.88
Road planings	8	2.1	6.0	75.0	1.57
Total:	32	8.4	17.0	53.1	4.47
Household					
Paper	6.6	1.73	0.345	5.23	0.090
Plastic	1.4	0.37	0.070	5.00	0.018
Textiles	0.8	0.21	0.200	25.0	0.052
Glass	2.0	0.52	0.240	12.0	0.061
Metals	1.5	0.39	0.25	16.7	0.066
Organics	7.7	2.02	0.0	0.0	0.0
Total:	20	5.2	1.105	5.53	0.29
Commercial	15	3.9	1.5	38.5	0.39
Total:	381.5		32.7		8.57

roughly 400 million tonnes per annum (DoE 1992b). Table 1 gives a breakdown of the general classes and types of waste produced. These figures are extracted from recent publications on waste and recycling (DoE 1991, DoE 1992b, DoE 1993a, ARUP 1991).

From this data a number of key points can be extracted:

1. A large proportion of 'waste' arisings are not controlled under any environmental legislation. Mine, quarry and agricultural wastes in total make up 41% of total waste arisings.

2. At the present, only 8.6% of all waste material is reclaimed. It could be argued that agricultural wastes, consisting mainly of slurry, are reclaimed by spreading on the land. However, I do not consider this to be an accurate assessment of fact, since spreading is primarily a disposal option, and does not take account of the wastes potential as an energy resource.

3. Although great effort is being directed by local authorities into meeting the Governments 25% domestic waste recycling target, even when all authorities achieve this target, the total reclaimed will only be 1.3% of all waste arisings.

4. Although waste such as dredgings and some sludges are disposed of in the sea, most materials are disposed of on land - at least 330 million tonnes annually.

Table 2 gives the figures for controlled waste disposal in the UK (DoE 1992b). The remainder of the annual waste arisings are disposed of by spreading on the land, by uncontrolled land-filling or land-forming, and by sea disposal.

The major route for waste disposal in the UK is landfill - 85% of controlled wastes are sent to landfill. Currently the UK co-disposes of many different types of waste in landfill, but this may soon change if new European directives are enforced.

Table 2. UK controlled waste disposal.

Destination.	Total, Mte/yr	% of waste	% of all arisings
Landfill sites	119.0	85	31.2
Incineration	5.6	4	1.5
Sea dumping	5.6	4	1.5
Other (recycling).	9.8	7	2.6
Total:	140.0		36.7

At the same time as potential changes to the type of waste disposed of, new guidance from the Department of the Environment (DoE 1993b) requires that landfill sites be constructed as 'total containment' vessels, similar to the American 'dry-tomb' designs. Until recently landfills were operated on the 'dilute and disperse' principle, where pollutants were allowed to leach into the environment in the hope that dilution or natural degradation would prevent any widespread environmental damage. When all these changes to the landfilling of wastes have taken place, it is likely that the costs of landfill could increase significantly.

4% of controlled waste is incinerated. Currently, most incineration is carried out in old plants primarily designed to achieve mass reduction. These plants 'pay' for themselves through savings in the total cost of landfilling waste, and some also generate electricity or process heat for industrial/ domestic use. Most of these plants will close shortly due to European directives on the operation of waste incineration plant because modification or retrofitting of pollution control equipment is too expensive.

Recycling is seen as a 'sustainable' alternative to landfilling waste, but the problem with recycling in the UK is that all the emphasis to date has been on domestic and commercial wastes, these being the most visible to the public at large. Other wastes which could be put to good use - colliery and quarry waste for example - have suffered as a result.

Recycling domestic waste has been promoted as the way forward to solving our waste and resources problems in the future. However, as highlighted earlier, even with 25% of all domestic being recycled, only 1.3% of all waste arisings will be reclaimed. Also, because the primary responsibility for waste recycling is being put on local authorities (the domestic waste collectors) rather than the waste producers, recycling is a very expensive option, especially in rural areas.

The other disposal option noted above is sea disposal - this is mainly for harbour dredgings and sewage sludge. New European agreements will stop the dumping of sewage sludge at sea from the end of this decade, so meaning that new disposal routes will have to be found. At this moment, the greatest consideration has been incineration. New agreements on the disposal of materials containing toxic ('Red List') substances may also curtail the dumping of harbour dredgings.

3. DEVELOPING A SUSTAINABLE WASTE MANAGEMENT SYSTEM.

Agenda 21, chapter 21, sets out a programme of measures to implement more sustainable waste management systems. The objectives and targets of the chapter are listed in table 3. Reading through, if all these measures were implemented in the required timescale, the next ten years will see a fundamental shift of emphasis in the waste management industry from a primary objective of waste disposal to one of reuse and recycling.

Thus far, the main thrust of national campaigns to solve the problem of waste has been towards the packaging of domestic goods. I consider this to be of little benefit to the sensible management of resources because at most, it represents less than 1% of the whole UK waste stream. Even with the latest recycling targets for industry declared by the Department of the Environment, the actual level of recycling may never exceed 5% to 10% of national waste arisings, because these targets are aimed at the packaging industry, rather than other areas such as sludge producers, or industrial waste producers.

The most recent criticism of recycling has come from the House of Lords Committee on the European Communities. Their report (H. of LORDS 1993) notes that the belief that recycling is 'morally right' is simplistic and unsound, and the "political drive to recycle anything and everything has become a matter of dogma unsupported by rational economic or environmental thinking".

One of the fundamental flaws with recycling is that it does not encourage minimisation of wastage. More efficient use of materials will be encouraged during manufacturing processes due to economic mechanisms, but the 'back end' of the system - the disposal of waste materials by society at large - is not encouraged. In some cases, even though the elimination of materials could produce an environmental benefit, many view 'recycling' as a more palatable alternative.

The Royal Commission on Environmental Pollution (RCEP 1993) noted that little information existed on waste minimisation, but recent reports (ENDS 1993a) show that the results of Government sponsored projects in Northwest England have been very successful. If we take the conclusions of the Royal Commission on Environmental Pollution, made in their Eleventh Report (RCEP 1985) and restated in their Seventeenth Report (RCEP 1993):

"The Commission's general approach to waste management can be presented as a four-stage decision procedure:
1st. wherever possible avoid creating wastes.
2nd. where wastes are unavoidable recycle them if possible.
3rd. where wastes cannot be recycled in the form of materials, recover energy from them.
4th. when the foregoing options have been exhausted, utilise the best practicable environmental option to dispose of wastes."

In their Seventeenth Report they go on to state:
"In terms of the four-stage decision procedure described at the beginning of this report, the first priority in waste management must always be to avoid creating wastes wherever possible. The view was expressed to us that waste minimisation if pursued with sufficient determination, might remove the necessity for incinerating any wastes, other than the most intractable clinical wastes. The reduction of waste arisings was discussed by the Commission in its Eleventh Report, which noted that avoiding waste can bring increased commercial benefit by altering a process, by changing from one process to another which might cost less in itself, or by decreasing disposal costs by generating less waste."

4. IS SUSTAINABLE WASTE MANAGEMENT POSSIBLE?

Although the Royal Commission echo the objectives of Agenda 21, I can see that it will be an uphill struggle to convince waste producers that investments in more expensive plants is necessary. The only way will be to begin to value waste for its true value in terms of finite resources and the energy invested in manufacture. Put more bluntly, the cost of waste disposal must rise to reflect the value of what we are throwing away.

To date, the main effort in terms of improving the sustainability of waste management systems has

Table 3. Agenda 21 waste management objectives and targets.

OBJECTIVES:	TARGETS:
* To stabilize or reduce the production of wastes destined for final disposal.	* By the year 1995 (in industrialized countries) to ensure that at least 50 per cent of all sewage, waste waters and solid wastes are treated or disposed of in conformity with national or international environmental and health quality guidelines. * By the year 2000, ensure sufficient national, regional and international capacity to access, process and monitor waste trend information and implement waste minimization policies.
* To strengthen procedures for assessing waste quantity and composition changes for the purpose of formulating operational waste minimization policies utilizing economic or other instruments to induce beneficial modifications of production and consumption patterns.	* By the year 2000, have in place in all industrialized countries programmes to stabilize or reduce, if practicable, production of wastes destined for final disposal, including per capita wastes (where this concept applies), at the level prevailing at that date. * To apply by the year 2000 programmes to reduce the production of agrichemical wastes, containers and packaging materials. * By the year 2000, to promote sufficient financial and technological capacities at the regional, national and local levels, as appropriate, to implement waste reuse and recycling policies and actions.
* To strengthen and increase national waste reuse and recycling systems.	* By the year 2000 (in industrialized countries) have a national programme, including, to the extent possible, targets for efficient waste reuse and recycling.
* To make available information, techniques and appropriate policy instruments to encourage and make operational waste reuse and recycling schemes.	* By the year 2000, establish waste treatment and disposal quality criteria, objectives and standards based on the nature and assimilative capacity of the receiving environment. * By the year 2000, establish sufficient capacity to undertake waste-related pollution impact monitoring and conduct regular surveillance, including epidemiological surveillance, where appropriate; * By the year 2025, dispose of all sewage, waste waters and solid wastes in conformity with national or international environmental quality guidelines.

been the expansion of recycling. There have also been some efforts to improve the knowledge available on waste minimisation. But the greatest discussion has been on the future for waste to energy plants - burning rubbish to produce electricity.

Recently, a developer seeking permission for a waste to energy plants (GEO 1993) stated that, "...there is a case by which the Energy Recycling Centre is a sustainable operation in its own right". It is difficult how this can be this position was arrived at; incineration plants produce waste ash which is contaminated with carcinogens. There is also the problem that the materials containing the greatest energy - plastics and paper/card - are precisely those which present the best options for recycling.

In energy terms, incineration makes no sense. Comparisons of landfill, recycling and waste-to-energy have recently been carried out (KIRKPATRICK 1992) and this shows that there is no distinct advantage for incineration over other forms of waste disposal when looked at in terms of energy balance. Also, if we take a traditional fossil fuels such as coal, it contains more energy per unit volume than waste. If we take data from ETSU (ETSU 1987) comparing coal to waste derived fuel (RDF), burning RDF to achieve the same energy output as coal requires 56% more material by weight - which equates to nearly three times the volume of RDF. Also, to produce the same amount of energy, burning RDF produces 420% more ash. This data is detailed in Table 4.

But, what is the answer to waste disposal? In essence, the purpose of waste management is to dispose of waste materials in a manner which causes the least damage to the environment. As shown by recent life-cycle analysis of waste disposal options (KIRKPATRICK 1992) the most

Table 4. A comparison of coal and RDF in energy production

Parameter.	Coal.	RDF.	Difference, percent:
Energy, GJ/te.	28.1	15.9	56%.
Density, kg/m3.	900	600	66%.
Ash content, %vol.	6.5	15.7	240%.

effective option for waste disposal is not a definite hierarchy of options. In terms of recycling the energy expended in reprocessing some materials may be as great as the energy used to manufacture them in the first place (in such situations, direct reuse or substitution would be advised).

If we relate each of the wastes noted earlier in table 1 to their 'best practicable environmental option' for 'waste management', the following options arise:

Firstly, there must be a greater emphasis on waste minimisation - over and above the current commitment given to recycling. This is required as part of the Agenda 21 programme, and will have to be addressed by the Government before submitting its review of Agenda 21 implementation to UNCED over the coming decade.

Secondly, there must be greater reuse of waste materials. Perhaps the most promising area for this is quarry/mining wastes. At the moment the only barrier to greater use is economic. The use of an 'aggregate tax' on primary extraction could even up this imbalance and encourage greater use of secondary aggregates, while at the same time preventing the destruction of the landscape through ever greater mineral extraction.

Thirdly, there is a place for recycling, but it must be noted that the most effective location for this responsibility is the waste producer. The only way to encourage manufacturers to consider the content/packaging and reprocessing of their products it to require that the waste product be returned to them for disposal - either directly or by contracting out to specialist recycling companies.

There are three other technologies, currently used in other countries, which could allow both cost effective management of waste materials, and reclamation of the materials or energy the wastes contain.

Firstly, the composting of organic wastes. This technology has proved promising for pure organic or 'garden' wastes, but there have been experiments recently involving the large scale composting of the organic fraction of domestic wastes (ENDS, 1993b) which have proven that it is a viable treatment method, although markets for the final compost have yet to be secured. The end product of this is a 'compost' which could be used for agriculture, or landscaping, but this would depend upon the the quality of the compost, and its pollutant content.

Secondly, anaerobic digestion. This has been used for sometime to treat agricultural wastes in Europe, but has yet to see its full potential in the UK. Utilisation of sewage sludge and agricultural slurry in digesters would not only remove some of the polluting potential of these wastes, it would also generate significant amounts of electrical power or methane gas.

Finally, pyrolysis. Although the principle of pyrolysis or gasification has been known for some time, it has yet to be widely applied to waste management problems. Plants have been designed to handle specific wastes such as plastics or rubber vehicle tyres, and these may soon be operated commercially. AEA Harwell and Herbert Bevan Ltd have developed a small scale plant (AEA-Bevan 1992) which can reclaim a large proportion of the mass of vehicle tyres, producing a flammable gas, an oil rich in useful aromatic hydrocarbons (with a calorific value equal to that of diesel), and an 'activated' carbon char. The design may soon be in commercial use across the UK.

5. CONCLUSION

The current structure of the waste management industry is un-sustainable, and efforts to increase recycling are disproportionately targeted at packaging wastes, rather than waste material as a whole.

To implement 'sustainable' waste management, it will be necessary to implement measures to require waste producers to take responsibility for the fate of their materials. It will also be necessary to encourage more sound use of materials by charging - through raw material taxes levied or waste disposal charges - values which reflect the value of materials in terms of the materials they contain or the energy involved in their manufacture.

Incineration, conversely to the image some are promoting, is not a 'sustainable' waste management option. For most wastes, with perhaps the exception of biologically hazardous wastes, other more effective solutions can be found.

The next ten years, if Agenda 21 is implemented effectively, will present the waste management industry with enormous challenges.

REFERENCES

AEA-BEVAN 1992. "MPD TP-2000 Tyre Recycler", AEA-Beven Ltd 1992, Building 404, Harwell Laboratory, Didcot, Oxon. OX11 0RA

ARUP 1991. "Occurrence and Utilisation of Mineral and Construction Wastes", Arup Economics and Planning, HMSO 1991.

DOE 1991. "Recycling", Waste Management Paper No.28, HMSO 1991.

DOE 1992a. "A Review of Options", Waste Management Paper No.1, HMSO 1992.

DOE 1992b. "The UK Environment", DOE/ Government Statistical Office, HMSO 1992.

DOE 1993a. "Digest of Environmental Protection and Water Statistics", No.15, HMSO 1993.

ENDS 1993a. "Yorkshire and Merseyside projects show the way on waste minimisation" p15, ENDS 221, June 1993.

ENDS 1993b. "Worm turns for composting", p15, ENDS 218, March 1993.

ETSU 1987. M.P. Landy, paper to Annual Wastes Management Conference, Torbay, 1987. Ref.: ETSU-L-20.

GEO 1993. "Environmental Statement on Proposed Energy Recycling Centre, Hespin Wood", GEO Ltd, 1992/1993.

H. of LORDS 1993. "Packaging and Packaging Waste", House of Lords Committee on the European communities, HMSO 1993.

KIRKPATRICK 1992. "Selecting a waste Management Option using Life-cycle analysis Approach", Packaging Technology and Science Vol. 6, 159-172, 1993.

RCEP 1985. "Managing Waste: Duty of Care", Royal Commission on Environmental Pollution 11th Report, Cmnd 9675, HMSO 1985.

RCEP 1993. "Incineration of Waste", Royal Commission on Environmental Pollution 17th Report, Cm 2181, HMSO 1993.

TCPA 1992. "Planning for a Sustainable Environment", Town and Country Planning Association, 1993, ISBN 1-85383-145-X; summarised in the 'Town and Country Planning Journal', May, Sept. and Oct. 1992.

UN 1987. "Our Common Future" (The Brundtland Report), World Commission on Environment and Development, 1987, Oxford University Press.

UNCED 1992. "Earth Summit '92", Regency Press 1992, ISBN 0-9520469-0-3.

Why Recycle?, Rainbow (ed.) © 1994 Balkema, Rotterdam, ISBN 90 5410 367 1

Recycling of trade waste in Wye

Richard T. Boden
Wye College, University of London, Ashford, UK

Abstract

Trade waste - from shops, restaurants, offices etc. - represents a significant proportion of total waste going to landfill. It has received little attention from Government compared to domestic waste, yet to the private waste management business it offers several advantages in terms of recycling over the domestic waste stream. This paper examines the success achieved in a village in Kent in diverting significant levels of trade waste from landfill.

Introduction.

Wye is a large village situated at the foot of the North Downs, on the river Stour between Ashford (4m) and Canterbury (8m). The resident population is just under 2,000, though this is swelled to over 2,500 for 30 weeks of the year by students at Wye College, University of London. The presence of the college, and the high spending power of its staff and students, explains the relatively large number of trade premises (45) for a village the size of Wye.

WyeCycle started a weekly kerbside collection from the 700 households in Wye in July 1989, which is now diverting 60 - 70% of total domestic refuse from the waste stream. In the summer of 1992 a collection of trade waste was launched, with the aim of establishing the cost-effectiveness of removing similarly high levels of waste as those achieved by the domestic collections. This report outlines WyeCycle's trade collections, and suggests how the project might be both taken forward in Wye and replicated in other villages.

Previous Disposal Methods.

Prior to WyeCycle starting their trade collection, an estimated 95% of trade waste went to landfill, with a small amount of cardboard and glass being recycled. From the 45 businesses in the village, disposal methods were as follows:

1,100 litre wheely bins	19 (23 bins)
"Green sacks" from L.A.	6
"Other"	20
TOTAL	45

In other words, it was either put out as domestic waste, burned or otherwise disposed of. Some businesses claim to produce no waste.

Costs.

WyeCycle put the financial cost of the above situation at £10,000/yr;

23 x 1100 litre wheely bins @ £6.52	£7,800
300 "green sacks" @ 65p	£ 200
Cost to firms (time, fuel etc) of other method	£1,000
Cost to L.A. of collecting and landfilling that put out as domestic waste	£1,000
TOTAL	£10,000

It is worth comparing this figure of £10,000 with the £8,250 spent collecting and landfilling the 275t of domestic refuse produced by the village each year. (This was 500t before WyeCycle started). Given that trade waste disposal costs are passed onto the consumer in the price of a product, people in Wye actually spend more on disposal of trade waste than they do on disposal of their own domestic waste.

Quantities

Given the varying amounts squeezed into a 1100 litre wheely bin, and the fact that some traders are reluctant to divulge how many sacks they put out as domestic waste (the official limit is two per week), we can only estimate the tonnage of trade waste produced each year. We put the figure at 3-4t per week, or 150-200t per year. This compares to 10t per week of domestic refuse.

Introductions of Recycling Collections.

All 45 trade premises were leafleted, informing them that WyeCycle would collect trade waste every Friday, free of charge, so long as it was separated into the following categories: 1) Cardboard 2) Paper 3) Glass 4) Organic.

Follow up visits were made to those premises which needed containers for organic waste (240 litre wheely bins), and most firms were visited over the first few weeks of the scheme to encourage participation and iron out any teething problems.

Participation Levels.

Twelve months after the introduction of trade collections, a total of 30 out of the 45 businesses are participating to a greater or lesser extent. Of these 30, 21 receive a weekly collection and 9 an occasional one. Of the 15 who do not take part, only 3 pay for their waste disposal, with the other 12 either producing no waste or falling into the "other" disposal category.

The following table gives a breakdown of the levels of participation in comparison to the previous methods of disposal employed:

USE OF WYECYCLE

Previous disposal methods	Weekly	Occasional	Never
Wheely bin	74%	5%	21%
Green sack	33%	33%	33%
Other	25%	25%	50%

This illustrates quite clearly that the level of financial saving to be gained by participating is a major factor in determining participation levels; the higher the cost of alternative disposal the greater the participation.

Providing a Complete service.

It soon became clear that some businesses were not receiving the level of financial savings they wished despite a high degree of participation in WyeCycle; even if we were taking three -quarters of their waste they were still having to pay for a wheely bin for "genuine" waste. To remedy this WyeCycle now provides a waste collection service alongside the recyclable collection; businesses may have a 240 litre wheely bin emptied for £1/week, with all recyclable waste still being removed free of charge. Thus there is a direct financial incentive; the more waste they separate, the less 240 litre wheely bins they need, the more they save.

This complete service has now been adopted by 10 of the 19 businesses that previously had a 1100 litre wheely bin, with the combined financial saving to these 10 firms being £55.20/week or £2,870/year.

Although firms probably cram their waste into our bins tighter than they did under their previous arrangement, the fact that most firms now only need one 240 litre bin instead of a 1100 litre one suggests we are recycling around 75% of the waste produced by these 10 premises.

Financial Aspects.

To collect from the 10 premises receiving the complete service takes one person one and a half hours in a Renault Master van. This comprises one hour actually collecting the materials and half a hour unloading the van and putting recyclables in their respective containers - colour separating the glass is a particularly time - consuming task.

The total cost of this is £10: £6 labour and £4 vehicle costs. This is covered by the £10 charged to service the 10 waste bins. Thus any income generated from the sale of materials is profit to WyeCycle. This surplus, given the current state of the markets, is minimal and the only way to turn the collection into a lucrative one would be to increase the £1/week charge for waste.

Future Development.

WyeCycle aims within the next 12 months to cut the total waste disposal bill for businesses in Wye to less than £5,000/year, half what is was before we started our service. (This will in effect be a greater than 50% saving, as the cost of traditional disposal methods is increasing rapidly as local tips close). To achieve this will mean recruiting more firms to our complete service, such that only five or so retain their 1100 litre wheely bin.

By achieving this 50% cut in costs, we will be simultaneously reaching a recycling rate for trade waste of 60 - 70%, the same as for domestic refuse in Wye.

Conclusions.

WyeCycle have established that trade waste is a major element in total waste created in Wye, in that the quantity produced is around one-third the total of domestic refuse and, perhaps more importantly, the financial cost of disposing of it is actually greater than that of disposing of domestic waste.

Businesses view the financial savings to be gained from recycling as a greater incentive than any environmental reasons, with those spending the most on disposal supporting WyeCycles collections more enthusiastically than those with low or no disposal costs.

The cost per tonne of collecting separated trade waste is less than half that of collecting source-separated domestic waste via a kerbside collection. This is due largely to the fact that each business yields a larger quantity of cleaner, homogeneous waste than each household, thus keeping labour costs down to one person compared to three or four employed on a domestic service. The financial incentive also means trade premises are generally more willing to separate waste properly, compared to households where we are relying largely on the "goodwill factor" to ensure proper separation.

Such a trade collection has hidden benefits to the local authority as they are collecting less trade waste "disguised" as domestic waste.

For further information about WyeCycle, please write to WyeCycle, Wye College, Wye, Ashford, Kent, TN25 5AH, U.K.

Why Recycle?, Rainbow (ed.) © 1994 Balkema, Rotterdam, ISBN 90 5410 367 1

Closed loop recycling – A Californian example

D.P. Ramsden
Evergreen Holdings Incorporated, Stockport, UK

ABSTRACT : The Californian model offers interesting insights into the factors which enable commercially viable closed loop recycling of automotive lubricating oil to take place. Evergreen companies have a unique combination of proprietary technology and integrated branding, from collection via rerefining and back to the motorist. This paper provides an overview of closed loop recycling in action, in California.

WHY RECYCLE ? WHY NOT ?

Taking the question posed by the title of this conference 'Why Recycle' as the basis for my paper, leads quickly to the riposte 'Why not' and perhaps more importantly for the green constituency 'How can we afford not to recycle'. The effective rejoinder to this has to be 'only if it pays'! It is essential for the health of any recycling activity that the environmental objective be supported by commercial viability. Unfortunately, the economic framework in which we operate often fails to give signals that accord with the desire to conserve resources and protect the environment. Since this a matter of political choice, it is usually possible to adjust the framework with the result that we can always ensure that 'it pays'. However, getting this adjustment to the economic framework correct, using legislation and social exhortation, is a subtle process and, as recent German experience shows, it is not easy to achieve the targetted outcome. Sometimes it is not even easy to identify the target outcome.

Today, I should like to present an example where it appears that desirable environmental objectives have been made congruent with commercial reward as a result of legislative adjustments affecting the economic framework. Considerable hard work was also required !

EVERGREEN HOLDINGS INCORPORATED

Evergreen Holdings Incorporated has been in the used oil management business in California for nearly a decade but only in the last few years has the considerable input of resource started to receive commercial success.

Through a number of separate operating companies Evergreen collects used oil, principally but not exclusively automotive material, as feedstock to its San Francisco rerefining plant. As described in detail later, the rerefining process which is one of only a small number worldwide to use modern hydrogenation technology, restores the oil to as good as, or better than, virgin quality product. This baseoil is then used to produce a range of automotive and industrial lubricants which are sold back into the marketplace, thereby closing the recycling loop. All steps of this process are commercially viable which, as mentioned before, makes this an example of the socio-economic framework supporting an environmental objective. Thus used oil is kept from undesirable disposal pathways whilst simultaneously conserving a resource for, potentially, multiple reuse.

DIMENSIONS

In the United States, an estimated 2.25 million tons of used oil is generated annually; Evergreen collects approximately 55 kt/a (2.5%) of this principally from the state of California but also Washington, Oregon, Arizona, and Nevada. The collected material yields the following products (approx.) :-

10% water - treated and discharged to sewer
13% fuel - used as fuel on site or sold **
12% asphalt flux - sold as bitumen fluxing agent
65% baseoil - sold for finished lubricants blending

** a proportion sold as gasoil

Apart from the water (which comes in with the feedstock), all product streams have commercial value; additional margin is obtained on part of the baseoil produced since it is blended and sold as Evergreen branded lubricant.

THE SOCIO-ECONOMIC FRAMEWORK

A number of factors now work together in California to make the collection of used oil for recycling to finished lubricant a commercially viable operation.

1 The Comprehensive Environmental Response, Compensation and Liability Act

Also known as CERCLA or Superfund. This federal legislation, enacted in 1980, has extensive and draconian penalties for remediation of contaminated sites. As long as a site remains in commercial operation, the Act is dormant; however, if the operator becomes bankrupt and abandons the site then all or any of the generators of waste found at that site will be held liable for whatever remediation costs are needed to restore it. A clean, professionally managed site which ceases operation will pose little risk whilst a site which has experienced escape of hazardous material resulting in contaminated soil or water will face extremely onerous remediation costs. Two US sites in Fresno and Salt Lake City are estimated to have incurred cleanup costs of $120 million. This has resulted in some substantial retroactive claims on companies who had thought that responsibility for their waste had been discharged by handing it to, apparently, responsible removal agencies.

This Act focusses the mind of generators with regard to auditing the disposal route for their waste and on the need to use a reputable and financially secure, fully permitted, bona fide recycler. It is, however, not possible to acquire indemnity from the pervasive reach of the act which is retroactive and almost unlimited in its power to seek out the deepest pockets.

2 Resource Conservation and Recovery Act 1976

This is the federal legislation which ensures 'cradle to grave' responsibility for hazardous waste by imposing management standards on generators and transporters of hazardous waste. Although used oil is not classified as a hazardous waste at the federal level, it must be managed as if it were in a number of states, including California. As with CERCLA legislation, this underpins a very positive approach to recycling.

The effect of the foregoing rigorous legislative framework has been to limit the number of participants involved in used oil collection and to progressively raise the used oil management standards for both generators and collectors. The impact of this legislation became fully felt in California in the second half of the 80's as the restrictive Hazardous and Solid Waste Amendments of 1984 came into effect.

This moved used oil collection from a product purchase basis to a waste disposal service; thus the oil value has shifted from being hydrocarbon price related to a service fee cost basis. This transforms the economics of collection !

3 California Oil Recycling Enhancement Act

A very recent addition to the aforementioned federal legislation is the above law which was enacted in January 1992; the first part, which requires manufacturers of luboil to pay a 16 c/USG levy to the California Integrated Waste Management Board, became effective in October 1992. The second part, which entitles fleet operators and certified collection programmes to receive 16c/USG for oil collected for recycling, has only functioned since April 1993. In this context, the luboil on which a levy is charged and a refund obtained is automotive product only; industrial oil is not part of this scheme. Clearly, most of the Evergreen story to date has developed prior to the operation of this Act and commercial viability has been achieved without this legislation. It is to be expected, however, that the effect of this law will be, as its title indicates, to enhance significantly the vailability of oil recovered from the DIY sector. It will achieve this by encouraging garages and workshops to accept used oil from the public since only by offering to take in oil will they be entitled to receive 16 c/USG for the oil generated from their own servicing activity. In addition, it is likely simply to increase the number of collection points available to the public. A minor, and possibly unintentional, side effect will be to make responsible DIY more attractive by 16 c/USG relative to having a vehicle serviced. Overall, a neat system for developing enhanced collection of used oil. It must be noted, however, that it requires a system of administration and that it does nothing to support closing of the recycling loop, as practised by Evergreen.

4 The Californian Psyche

In the State of excess, conscience is salved by concern for the environment but only at the margin ie it is a second order effect. Thus using cars with relatively large engines and changing your oil every 3000 miles (as many do) is offset by ensuring that the used oil is managed in an environmentally responsible way. As a slight digression, a similar phenomenon is apparent in the resource being poured into the Low and Zero Emission Vehicle projects - the names only being applicable based on emissions released at point/time of use. Equally, how environmentally friendly is the annual model change. Since voluntary, fundamental attitude change can be very slow or nonexistent when the changes are perceived as lifestyle constraining, the important thing is to work on the possible. In California, this is where Evergreen is able to return a modest portion of the used oil stream back to marketable lubricant. There is also no doubt that the environmental lobby is active, articulate and relatively well supported in America with California having some of its strongest proponents. Environmental issues gain detailed coverage, particularly in the press. Philosophical ones less so.

THE HAZARD LEVEL

Over many years, considerable debate has gone into the subject of the very real hazards associated with the management of post consumer used oil. A number of extraneous hazardous products regularly appear in used oil as a result of improper disposal, not of the oil but of that particular product; it is merely the fact that oil is an opaque liquid into which other difficult-to-dispose-of liquid wastes are 'lost'. This is an issue relating to the management of hazardous liquid waste in general rather than used oil in particular; however, the problem remains. There is no doubt that, when improperly discarded, used oil can cause considerable ecological damage. Even when disposed of in accord with conventionally accepted criteria, it must be recognised that it has the potential to introduce a number of hazardous species into the atmosphere eg heavy metals and post-combustion chlorinated species. The following list shows the harmful products often associated with as-collected used oil:-

.1 Lead - even where the gasoline pool is substantially lead free, lead derived from bearing wear is found in used oil at potentially harmful levels. (as recently as March 93, California EPA expressed concern about lead release from used oil, amongst other sources).

.2 PNA/PCA/PAH - by whatever abbreviation, these potentially carcinogenic components are found in harmful levels.

.3 Metals - zinc, chromium, cadmium, vanadium, and copper present from additive package or engine wear.

.4 Solvents - present as a result of improper disposal of solvent or from parts deoiling in workshops.

.5 Chlorine - as .4, chlorinated products improperly discarded.

.6 PCB - only very occasionally found now, from period of use as an insulating oil.

.7 Hydrocarbons - light components from internal combustion engine or contaminated fuel disposal.

.8 Particulates - combustion products and sediment.

PATHWAYS FOR DISPOSAL

.1 Any of the aforementioned plus the luboil hydrocarbon can find its way into soil and/or water courses if improperly released to the ground. A small amount in water courses can render large volumes non-potable.

.2 Combustion will destroy some of the above

species but the extent can be uncertain under all but closely controlled operations eg in incinerators and possibly cement kilns. Some components eg lead and metals, will still be released to atmosphere after the combustion process. In California, restrictions on the lead levels permitted in fuels mean that much of the used luboil collected has to be routed to ships bunker fuel. Thus the hazard is removed from the shores of the state.

.3 Recycling offers the most attractive route to managing the above contaminants. The luboil is recovered for reuse. PCA (and potentially PCB) are destroyed with greater than 99% efficiency. Organic chlorine is converted to HCl and neutralised. Metals and very heavy hydrocarbons leave as an asphaltic residue stream which is used as a flux for bitumen manufacture where they become locked into a non-leachable solid matrix.

.4 Whilst some of the foregoing benefits can be obtained by the gasification/coking option, clearly the valuable luboil molecules are destroyed in this process.

THE EVERGREEN OPERATION

Evergreen Holdings has grown both organically and by acquisition; essentially, its operation has three components namely collection, rerefining, and sales/marketing. It is also involved in vacuum truck activities which are an additional source of oil/water feedstock into the plant, together with the recycling of used antifreeze and the collection of used oil filters.

The plant was originally built within the constraints of a 10 M$ financing bond from the state of California with the result that it was relatively small compared to the potential market. Active debottlenecking has, however, resulted in a considerable increase over original design capacity (from approximately 10 to 15 million USG/annum) although an upper limit, set by the current operating permits, is now coming into sight. Energy consumption per tonne of finished baseoil is some 60% of that for virgin production whilst capital costs also compare favourably; both important factors when looking for additional baseoil manufacturing capacity.

Collection

A fleet of more than 65 trucks is operated by Evergreen in its collection business throughout California and adjoining states. Collections are made from over 6000 sites including auto workshops and fleet maintenance depots (75%) as well as DIY, government and industrial locations. Most drivers have well established customer bases thus building confidence in the quality of the material being routinely collected. In addition, a minor amount of product is brought in by small, independent collectors. All used oil undergoes a 10 minute test programme before being accepted for discharge from the collection vehicle; sealed samples are retained from each pickup point in order to provide an audit trail in the event of undesirable components being detected. On well established routes, the collection service can operate almost independently of the generator; the driver will call in on a regular basis and uplift the used oil whenever a sufficient quantity has accumulated. In-cab computers have recently been installed to create instantaneous invoices showing the collection charge and the quantity of oil removed.

REREFINING

Used oil undergoes six processing steps in the Evergreen plant with hydrogenation as the key stage in restoring the oil to a quality level equal to virgin material.

Step 1 - a proprietary chemical pretreatment step which a) dramatically reduces fouling throughout the plant and b) greatly increases hydrotreating catalyst ife. Prior to developing this detailed understanding of the phsysico-chemical processes occurring in the plant, downtime amounted to well over 100 days per annum compared to less than 20 now.

Step 2 - after filtration and product heat exchange, the feed stream passes via a hot oil exchanger into an atmospheric flash column where water and gasoline type hydrocarbons are flashed off at about 145 C.

Step 3 - in the next stage, after hot oil preheat heavier diesel type components are removed under vacuum flashing conditions (260 C and 0.08 atmosphere).

Step 4 - the remaining hydrocarbon stream containing dirt, metals and polymeric material is then fed to a high vacuum, thin film evaporator (LUWA type) which provides high temperature, low residence time flash conditions from which a residual bottom stream is produced that is used as a flux in the asphalt industry. (370 C and 0.01 atmosphere).

Step 5 - high temperature catalytic hydrogenation in three reactor stages then removes sulphur, chlorine, and, most importantly, any remaining carcinogenic species from the luboil. This key processing step also improves product colour and odour (final reactor exit conditions 340 C and over 80 atmospheres).

Step 6 - The product is then split into 100 SN and 300 SN fractions (required by the blender market) in a conventional fractionation column. Yields are very stable reflecting consistency in the feedstock collection system and quality is indistinguishable from virgin product.

The plant is located in an environmentally sensitive area at the southern tip of San Francisco Bay. Apart from having to be a good neighbour to nearby residential property, it operates under an extremely rigorous permitting scrutiny maintained by the state and city authorities. It is the only fully permitted plant ie inspected from new, of its kind in California; other collectors generally only having licences by virtue of longstanding operations on their sites.

SALES/MARKETING

In its early days, the Evergreen plant delivered very erratic performance principally due to the fouling that occurred in unexpected amounts in unexpected places. Unforeseen chemical reactions were occurring in addition to physical separation processes. This led to variability in product quantity and quality which clearly had to be reflected in a discounted price. This merely added to the image problem which has long plagued recycled luboil produced by generally unsatisfactory "recycling" practices. Thus the new era of hydrogenated, quality, refined product was delayed somewhat whilst these prolonged teething trouble were resolved by the introduction of proprietary Mohawk pretreatment technology.

During this period, baseoil produced by the plant was sold, at a discount to virgin product, to blenders who used it in formulated lubricants but without declaring its origin. Now, with dramatically improved plant performance and a changed social attitude towards recycling in general, baseoil is sold virtually at parity with virgin material. In addition, formulated lubricants bearing the Evergreen name have been on sale in California for well over a year, proudly indicating their Environmental origin. Grades produced include automotive, ranging from 5W/30 to 20W/50, Automatic Transmission Fluid (soon to obtain Dexron III approval), and hydraulic grades (ISO 32/46/68). A number of major US lubricants companies (Chevron, Unocal and Arco) have con tracts to take Evergreen recycled baseoil for blending into their own branded Environmental product ranges. It is worth noting, also, that in Canada Shell, Exxon and Chevron have been selling lubricants containing rerefined components for several years.

The lighter hydrocarbon stream resulting from the defuelling steps is used within the plant as fuel for process heating whilst some of the heavier distillate stream is sold as gasoil. The residual stream containing high molecular weight hydrocarbons and metallic compounds has a viscosity that makes it suitable for use as a bitumen flux. Thus all the material collected for recycling achieves a commercial value for the second time in its life and in the case of the luboil molecules possibly more times than that.

CONCLUSIONS

With its proprietary process technology, Evergreen provides an example of commercially viable closed loop recycling in practice. For the future, plans are well advanced for building a new refinery in Arizona with three times the capacity of the San Francisco plant; in preparation for this, collection capacity is growing in the Los Angeles area. Construction has recently started in Indonesia on a plant designed by CEP, an affiliated process engineering company, whilst plans for expansion into Europe are also under development. A growing public brand awareness and the growth of environmentally friendly products in general should secure a green future for Evergreen.

Why Recycle?, Rainbow (ed.) © 1994 Balkema, Rotterdam, ISBN 90 5410 367 1

Optimising office wastepaper collection schemes to produce high quality fibre

Kate Cathie
Pira International, Leatherhead, UK

ABSTRACT: Utilisation of secondary fibre in the production of printing and writing papers and high quality tissue products in the UK is very small. As a result of this and our low indigenous wood production, approximately 1.6m tonnes of virgin pulp are imported into the UK each year contributing substantially to the balance of payments deficit.

One of the reasons for the lack of waste-based printing and writing and high quality tissue production in the UK has historically been a lack of suitable fibre. This situation is now changing as it is recognised that office waste, which was previously an untapped resource, can be suitable for the economic production of these paper grades provided that care is taken in its collection and processing.

This paper provides figures on the potential tonnages of office wastepaper in the UK based on employee data and paper yield figures from collection schemes.

Wastepaper processing stages are briefly described in relation to office waste utilisation. Results from laboratory studies undertaken to evaluate the fibre quality of office wastepapers are discussed and compared with those of an equivalent virgin fibre.

Finally, the effects of possible future changes in printing processes and office practices on the viability of office wastepaper collection are given.

1 INTRODUCTION

Recovered office wastepaper is generally most suitable for usage in the manufacture of printing and writing grades and high quality tissue products. Printing and writing production in the UK is currently predominantly based on virgin fibre, much of which is imported as chemical pulp. Only 7% of the fibre used to manufacture the 1,478,000 tonnes of printing and writing papers in the UK in 1991 was wastepaper. In contrast, tissue production has a higher wastepaper utilisation rate of about 40%. However, much of the wastepaper used is lower quality and is incorporated into lower standard products such as towelling and industrial wipes.

Historically, the low utilisation of wastepaper in printing and writing grades and high quality tissue products has been due to problems in the supply of suitable wastepaper grades and a belief that the technology required to produce a suitable pulp from wastepaper was not available. Although both of these aspects have now to a certain extent been resolved, there are still imbalances in the amount and types of wastepaper collected and the amounts and types of wastepaper required by mills. The problem is simple; people will not collect wastepaper unless there is an end-use for the material and mills will not invest in new equipment to produce wastepaper unless they can be ensured of a steady and suitable fibre supply.

The aim of this paper therefore is to demonstrate how much office wastepaper is potentially available for collection, highlight the processing that it will require and the quality of the fibre that can be obtained. Predictions as to how future changes in office practices and printing technologies will impact office wastepaper usage are also given.

2 POTENTIAL FOR OFFICE WASTEPAPER RECOVERY

The amount of office wastepaper which can be recovered will depend on a number of factors including:

- the amount of paper consumed
- the number of office employees involved
- the yield obtained from each employee

Yields of paper collected from individual establishments will obviously depend on the office type and the types of paper collected. In a scheme operated by Pira International, the total wastepaper yield per office occupant per week was found to be 2.4 kg, whilst the yield of white wastepaper only, often deemed to be of higher quality, was 1.4 kg per occupant per week. Using these figures and employee data for office based employment (Table 1), the potential for office wastepaper recovery in the UK can be estimated.

Although these calculations do not take into account paper storage and paper clear-outs etc, they do show that over 770,000 tonnes of wastepaper per annum could be recovered from these employment sectors. Over 453,000 tonnes of this paper could also be higher quality white paper, which may require less processing (see Section 3).

The effect on the UK trade deficit due to the import of virgin pulp and paper products through the collection of even half of this white material, would be dramatic. Using these figures, imports of pulp could be reduced by about 200,000 tonnes per annum at a value of over £69 million.

Table 1 Amount of office wastepaper which could potentially be collected in the UK

	Employment type			
	Banking finance	Public administration	Education	Total
Full-time employees in June 1992 ('000s)	2,583	1,811	1,836	6,230
Amount of wastepaper which could potentially be collected (tonnes)	322,358	226,013	229,133	777,504
Amount of white wastepaper which could potentially be collected (tonnes)	188,042	131,841	133,661	453,544

Additionally, if the wastepaper was converted into printing and writing papers and tissue products suitable for general office use, imports would be further reduced.

3 PROCESSING OF OFFICE WASTEPAPER

Irrespective of the type of wastepaper, it cannot be used in the manufacture of paper and board products without undergoing some kind of processing. The intensity of processing and the stages required depend on the wastepaper type being utilised and the product into which it is being made. The goal of all wastepaper treatment plants is to maximise the recovery of re-usable fibres and minimise the amount of residual foreign impurities present. Even for a comparatively clean wastepaper type, such as sorted office wastepaper, many impurities such as inks, staples, adhesives etc will undoubtedly be present which have to be removed. Generally, individual stages of wastepaper processing equipment remove or mask different types of impurities and like most separation stages, none are 100% efficient. A multistage approach is therefore required.

In order to produce re-usable fibre from office wastepaper, the following techniques are usually employed:

- pulping
- screening
- centrifugal cleaning
- deinking
- dispersion
- bleaching.

The first four stages are designed to remove contaminants; the latter to make them less visible and problematic. Each of these stages will be described briefly in turn with specific emphasis on their role in the processing of office wastepaper.

3.1 Pulping

The first stage in any wastepaper stock preparation plant is pulping. Stock treatment in the pulper is designed to maintain as much difference as possible between the physical properties of the contaminants and the fibres, whilst dispersing the fibres. As most office wastepaper will be printed, the pulping stage also plays a vital role in ensuring that ink is released from the fibres in the correct form and that it does not redeposit back onto the fibres.

Office wastepaper, like most wastepaper types, will contain many adhesives originating from labels, envelopes etc. Such adhesives are the main cause of tacky particles in wastepaper-using mills, known generically as stickies. As stickies cause problems all the way through wastepaper use, from processing, to papermaking, to converting, many stages in stock preparation are designed with specific emphasis on their removal, including pulping.

Many mills processing this wastepaper type will therefore use high consistency pulping (15-18%), to keep the size of contaminants, primarily stickies, to a maximum, enhancing their removal by subsequent processing stages. As a bonus, energy and chemical usage are also reduced.

3.2 Screening

Screening is a mechanical separation technique which relies on exploiting the physical size difference between a fibre and the contaminant to be removed. Usually, two forms are employed; high consistency coarse screening which takes place at the beginning of stock preparation to remove large impurities and low consistency fine screening, which is carried out further downstream to remove smaller contaminants such as stickies, fibre bundles, ink flakes etc. Pressure screens utilised in wastepaper-using mills are usually pumped through machines whose functional element is a plate or cylinder perforated with slots or holes. Accepted pulp passes through the openings, but larger debris is held back and is guided to a reject outlet.

3.3 Centrifugal cleaning

Centrifugal cleaning is used to separate contaminants with a significant density difference, either higher or lower, than the fibre and water suspension. Like screening, centrifugal cleaning is used at different stages of stock preparation. High density centrifugal cleaners are used to remove very dense contaminants such as staples, paper clips etc, early in the process and low consistency centrifugal cleaners are used later in the system to remove smaller, less dense contaminants such as stickies, ink flakes etc. Cleaners used for the different purposes differ in design. However, in all high density cleaners, cleaner function is based on the vortex flow of pulp which causes the heavy particles to migrate to the inside of the vortex and then be rejected through reject outlets.

3.4 Deinking

In order to produce high grade products, such as tissues and printing and writing papers from printed office wastepaper, the ink has to be removed through processes known aptly as deinking. The objective of all deinking plants is therefore to improve pulp brightness and reduce speckiness. Traditionally two separate techniques have been used to remove ink; washing and flotation deinking. Each is discussed briefly in turn.

3.4.1 Washing

Washing is a hydraulic separation process, whereby water is removed from the slurry, carrying with it ink particles. Basically, the stock containing fibre and ink is successively diluted and thickened. At each thickening stage ink and other small particles are removed leaving fibre behind with the remaining ink. Effective washing is dependent on having small sized particles, less than 10-15 μm, although particles between 0.5 to 5 μm are the most efficiently removed (Figure 1).

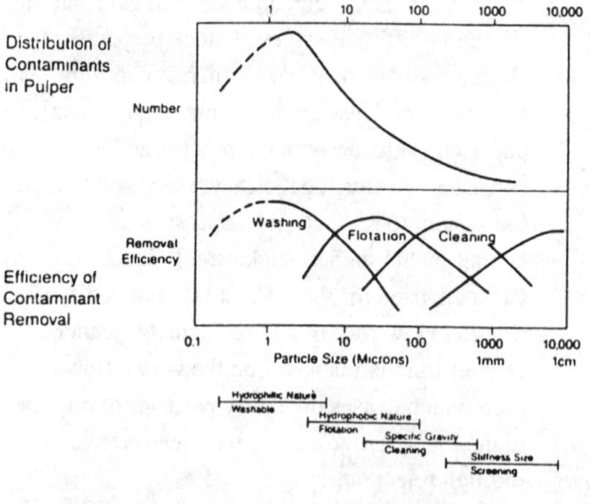

Source: "Recycling and separation technology", P LeBlanc and M McCool, Copyright TAPPI 1988

Figure 1 Particle size and removal efficiency

3.4.2 Flotation

Flotation is a chemi-mechanical process based on the phenomenon that if air is bubbled through a water suspension of particles, the bubbles will preferentially adhere to those particles not wetted by water. Ink particles present in the pulp are therefore conveyed to the surface of the suspension on air bubbles and removed as a froth. Ink removal efficiency, as with washing, depends on particle size (Figure 1). Particles in the size range of 10-100 μm are effectively removed, but efficiency drops with increasing particle size.

Mills are now commonly using a combination of the two deinking processes realising the benefits of both. With an office wastepaper furnish this is a particularly valid approach as many of the ink types present will be toner based from non-impact printing processes (NIP); laser printers and/or photocopiers. These inks, unlike those from offset or gravure printing, are not easily dispersed during the pulping stage. As a result, many may be too large to be removed by washing and some may even be too large to be removed by flotation. Not all NIP papers are however, difficult to deink; some can be easily deinked by flotation (see Section 5).

Ink removal is not the only criterion which needs to be considered when deciding which deinking techniques to install. Mills manufacturing tissue from office wastepaper will require a washing stage to remove ash (inorganic material) from the pulp, as this decreases tissue quality.

3.5 Dispersion

Dispersion is a system which employs intense shear forces to reduce contaminants to a size which minimises their effect on product quality or machine runnability. In this process pulp is forced through discs with interlocking teeth, generally at high temperature, 70 to 120°C and high consistency, 20 to 40%.

Traditionally dispersion has been carried out at the end of wastepaper stock preparation as a last resort to mask the presence of any remaining contaminants. Now and perhaps more so in the future, its ability to reduce the size of ink particles may well alter its role and position in stock preparation systems. As mentioned in the previous section, many of the toner based inks found in office waste produce particles too large to be removed by conventional flotation and wash deinking techniques. Use of a dispersion stage prior to deinking or prior to an extra deinking stage has proved advantageous with this wastepaper type.

3.6 Bleaching

Brightness is one of the most important parameters in the manufacture of high quality tissue products and printing and writing papers. Thus in addition to efficient cleaning and deinking, office wastepaper also frequently has to be bleached (more correctly termed brightened) and colour stripped to enable its re-use. The most common bleaching chemicals used are hydrogen peroxide, an oxidising chemical and sodium hydrosulphite, a reducing chemical. Formamidine sulphinic acid (FAS), another reducing chemical is also rapidly gaining favour. Sodium hypochlorite was widely used as a bleach for furnishes such as office wastepaper, but environmental issues surrounding the use of chlorine- based compounds and the movement away from chlorine usage by the virgin fibre industry has now caused the selection of other bleaches for this purpose, eg ozone.

4 QUALITY OF RECOVERED OFFICE WASTEPAPER

Many laboratory and pilot plant studies have been carried out to assess the quality of fibre obtained from office wastepaper collection schemes. One such trial involved wastepaper collected from several civil service offices in the UK. This wastepaper was found to contain on average 35-40% bleached softwood pulp and 60-65% bleached hardwood pulp. Its strength properties, paper machine runnability properties and visual/optical properties were therefore assessed and compared with those obtained from laboratory made virgin fibre-based paper containing 35% bleached softwood fibre and 65% bleached hardwood fibre. The wastepaper sample was deinked by flotation before these evaluations were made to provide a more realistic comparison.

Results from each set of properties will be discussed in turn.

4.1 Strength properties

Strength of the wastepaper pulp was found to be superior to the virgin pulp (Table 2).

Table 2 Strength properties of the wastepaper pulp were superior

Strength property	Wastepaper pulp	Virgin pulp
Burst index ($kPam^2g^{-1}$)	3.01	1.20
Tear index (mNm^2g^{-1})	8.97	8.70
Tensile index (Nmg^{-1})	47.3	28.0

These differences are presumably due to the past treatment of the wastepaper furnish.

4.2 Paper machine runnability properties

Paper machine runnability properties of both pulps are given in Table 3

Canadian standard freeness, CSF (which gives a measure of the drainability of pulps), although lower for the waste-based pulp, would be

Table 3 Paper machine runnability properties of the wastepaper pulp were good

Property	Wastepaper pulp	Virgin pulp
CSF (ml)	450	575
Ash content (%)	3.7	<1.0
Contaminant level (%)	0.6	0
Stickies per 100g OD fibre	0	0

Table 4 Visual and optical properties are acceptable

Property	Wastepaper pulp	Virgin pulp
ISO Brightness (%)	77.1	88.0
Dirt count (mm^2m^{-2})	2.5	0.5

perfectly acceptable for printing and writing and tissue production.

Ash level was reduced to an acceptable level by the deinking process to enable the pulp's usage in tissue products. This level would also be acceptable for printing and writing manufacture.

Contaminant level of the waste-derived pulp was low and mainly consisted of staples and paper clips. In an industrial situation these would easily be removed by centrifugal cleaners during stock preparation.

As already implied, the presence of stickies in many wastepaper grades is probably the predominant factor preventing increased usage. Removal of stickies can be very difficult. No stickies were found in this wastepaper pulp. This is excellent and shows how the quality of wastepaper can be improved by diligent source sorting.

4.3 Visual and optical properties

Resultant brightness after flotation was lower for the wastepaper pulp than for the virgin pulp used for comparative purposes (Table 4). However, in an industrial system, further brightening stages such as bleaching would also be used. Typical brightness levels for printing and writings and tissue grades are 80-85%. It is anticipated that this pulp would easily reach this target by final stage bleaching.

Stock speckiness of the wastepaper pulp after deinking (measured by dirt count) was higher than that of the virgin pulp, but would be acceptable. As mentioned earlier (see Section 3.5), improvements in speckiness could be made by combining wash and flotation deinking or by disperser treatment. This latter process would reduce the visual impact of ink specks and could improve their removal by a post flotation stage.

Resultant speckiness after deinking was in fact surprisingly low for a furnish of this type. This may be due to the presence of a low amount of non-impact printed (NIP) papers or the presence of NIP papers which can be readily deinked (see Section 5).

5 FUTURE IMPACTS ON OFFICE WASTEPAPER COLLECTION AND RE-USE

Several years ago there was talk of a paperless office, due primarily to improved computer technology. However, in direct contrast to this, the use of more word processors and personal computers has led to the increased use of paper in offices. There is no doubt that this trend will continue into the future and that the greatest change to office wastepaper furnishes will be an increase in the amount of NIP papers present. Growth in consumption of laser printed paper is expected to increase by 8-10% per annum until

1995(1). Growth in the consumption of xerographic papers in Western Europe has increased annually by 20% over the past 15 years and is expected to increase in the future at a rate of 10% per annum(2). A recent survey of Pira International's office waste showed that 50% of it is NIP.

These changes in theory should make office wastepaper more difficult to deink. However, there are great differences in the deinkability of different NIP papers. For so far unexplained reasons, some NIP papers deink far better than others, presumably due to differences in the toner formulations used, printers used, basepapers used, or even the age of the print(3,4,5,6,7). This effect may become even more complicated in the future as it is predicted that the diversity of imaging techniques used for NIP will increase, which may also influence deinkability.

Further expected developments are that colour printers will supersede their black and white counterparts and that printers designed to produce more dots per inch, to improve print quality, will be used.

The effect of either of these changes on deinkability are at present uncertain, although logically the use of more toner on the paper can only worsen the situation. Much work is however being carried out by wastepaper equipment suppliers, chemical suppliers and research associations, to find new ways of processing papers with high NIP contents and several methods have already been suggested(8,9).

Looking at the other main contaminant of office wastepaper, adhesives, the situation seems more hopeful. Many adhesive companies are currently seeking ways to reduce the impact of their products in wastepaper-using mills, or ways to produce adhesives with characteristics which enhance their removal by conventional equipment.

6 CONCLUSIONS

This paper has demonstrated that over 777,000 tonnes of office wastepaper is potentially available for collection in the UK, enough to supply 3 to 4 large papermills.

Processing of such wastepaper is definitely technically feasible, although admittedly not easy. Problems may be encountered with large volumes of difficult to deink NIP papers and contamination by adhesives. However, these problems are not insurmountable. The problems of adhesives can be reduced dramatically by careful source-sorting. Close liaison with the end-user of the wastepaper or wastepaper merchant should ensure that the wastepaper supplied is of the quality desired.

In the future further challenges will be faced by the wastepaper users, primarily because they are dependent on a raw material which is not designed for re-use but obviously and rightly for its original purpose.

What is needed is for both issues to be considered as new developments are made. Again communication is required.

7 REFERENCES

1 Print World, Vol 22, No. 1, October 1986, p36

2 *Paper Requirements for Non Impact Printing*, Pira project PB/C/87/5

3 Okada E, Urushiata H, *Deinking of toner printed paper*, Pulp and Paper, April 1991, p39

4 Carr W, *The state of the art of deinking difficult inks*, Proceedings of the Tappi Pulping Conference, 1991

5 Cathie K, Crow H, *Difficult to deink wastepaper grades - the way forward*, Proceedings of the Tappi Pulping Conference, 1991

6 Cathie K, Fabris I, Fullick C, *Influences of ageing on the deinkability of different non-impact printed grades*, PTS-PTI Deinking Symposium, Germany 1992

7 Cathie K, Moore G, Burdett S, *Understanding why laser printed papers deink differently*, Pira Conference - Recent Developments in Wastepaper Technology, March 1993

8 Longhini D, Hipolit K, *Agglomeration and removal of electrostatic inks from post consumer wastes*, Proceedings of the Tappi Papermakers' Conference, 1992

9 Olson C, Sutman F, Richmann S, Letscher M, *Deinking of laser-printed stock using chemical densification and forward cleaning*, Proceedings of the Tappi Papermakers Conference, 1992

Why Recycle?, Rainbow (ed.) © 1994 Balkema, Rotterdam, ISBN 90 5410 367 1

Industrial lubricating oil – The environmental challenge
A unique opportunity to be green and yet save money

Mick Bradley
Lanstar, Halesowen, UK

ABSTRACT

The amount of waste oil generated in the UK and efficient means of best using this important resource are reviewed.

INTRODUCTION

Oil, a non renewable resource, is one of the major raw materials used in the Western World. Current forecasts suggest that there are sufficient oil reserves to last for only a further 30-50 years.

Doubtless new oil fields will be found, and improved oil field technology will be discovered which will extend this resource.

Nevertheless, the theme of this paper is to review the best means of husbanding this vital resource.

BASIC CONCEPT

The basic concept of any of these processes is that oil molecules, be they aromatic, naphthenic or paraffinic, are intrinsically extremely stable. They have, after all, been formed by the effect of heat and pressure on organic matter over the course of millions of years underground.

These oil entities of various molecular weights are the building blocks of automotive and industrial lubricants. Their properties being further enhanced by the presence of additives and form the range of mineral, automotive and industrial lubricants available today.

After use these oils can often be in an apparently dreadful state. A quick visual examination is sufficient to confirm this. Yet initial reactions can be misleading. This emulsion is mixture of oil, water, solids, additives and metal fines with a wide range of decomposition products.

Importantly the original oil molecules are still present in their original state and, if they can be successfully segregated, can be successfully recovered.

AUTOMOTIVE AND INDUSTRIAL OILS (MARKET SIZE)

U.K. industry consumes over 700,000 tonnes per annum of automotive and industrial lubricating oils. Primarily as hydraulic, gear, metalworking, automotive and lubricating fluids. These 700,000 tonnes per annum of product ultimately end their life as a residue.

These oil residues, if not correctly recovered or disposed of, represent a major environmental threat. Yet today over 300,000 tonnes per annum of oil (that is over 40% of the total available), still disappear into the environment.

The remaining 400,000 tonnes per annum of oil are regenerated by one of three routes:

(1) oil recovery
(2) oil recycling
(3) distillation.

DEFINITION OF TERMS

1) Recovery That is to transform the used oil into a secondary support fuel. In this area I would include the handling of flammable solvents in cement kilns.

2) Recycling That is to reprocess and return an oil for reuse in its original application. (Original performance characteristics).

3) Distillation To distil an oil and collect various base oil fractions for reuse.

OIL RECOVERY

A large volume of used oil arises as a mixture of many oil types.

This is most commonly handled by a simple but efficient process known as oil recovery. This involves dewatering, removing particulate matter by means of a centrifuge. If necessary the viscosity and calorific value are adjusted. The feedstock and final oil blends are carefully tested to ensure that pcb and heavy metal contents fall within levels proscribed by the authorities.

Interestingly one major environmental advantage of secondary fuel oil over virgin product is its low sulphur content. This recycled oil is used as a secondary support fuel in selected industries.

400,000 tonnes per annum of oil is recovered by this route.

This recovery option represents the best outlet for certain oils, such as soluble cutting fluids or automotive lubricants. Yet a fuel is a once off use.

The same philosophy is used when flammable solvent mixtures are recovered and used as support fuels in cement kilns.

OIL RECYCLING

In my view, a far better option exists, and that is the recycling option where the oil is returned for reuse in its original application. This is surely the best practical environmental option.

In the U.K. today, only 35,000 tonnes per annum of oil is recycled. We believe that if oils were correctly collected, segregated and stored, it would be possible to recycle up to a further 200,000 tonnes per annum of industrial lubricating oils.

The main advantages of oil recycling are that:

1) An important natural resource is recovered for reuse. The oil molecule is, after all, intrinsically extremely stable. It can be recycled again and again.

2) It is cost effective in that the cost of using a recycled oil is far less than that of a new lubricant. Typically savings of up to 50% can be made.

3) It is technologically sound. Our experience based on 35 years of oil recycling shows that the recycled oil performs in even the most severe applications.

This process depends upon a strict segregation of all feedstock. Thus the customer can always operate in the confidence that it is his own oil which is being returned for reuse. It also depends upon stringent in-line process controls, which can only be provided with a modern, fully equipped Technical Department staffed by experienced chemists.

This guarantees that the customer is assured of oil being returned to an agreed specification.

PROCESS DETAILS

The used oil is recycled by a process of dehydration, filtration, additive replenishment and viscosity adjustment.

Once initial laboratory evaluation has shown that a used oil is recyclable, the used oil is collected by a dedicated tanker fleet using multi-pot vehicles or in 205 litres barrels. These are evaluated on arrival and then placed in segregated discreet tanks. Quantities from as little as 2000 litres can be handled, up to large systems of over 200,000 litres.

Oils are then charged to one of our process vessels where it is heated by indirect gas burners to remove water and is then further heated to sterilise the used oil. At the same time chemicals are charged to the process vessel in order to lighten the colour of the used oil and to remove partially decomposed additives.

Once approved by stringent in-line q.c. tests, the oil is filtered on a plate and frame filter press. Again technical approval is obtained prior to releasing oils for the final stage of the operation, which is blending with base oils to adjust viscosity and treatment with state of the art additives, which ensure that the recycled oil is fit for the most arduous application. This final product, after testing, is held in bulk storage awaiting return to the customer. Final sampling occurs halfway through discharge of product onto the road tanker. This retained reference sample is held for a period of three months.

DISTILLATION

The distillation of base oils from a mixed oil feedstock has been known for many years and the technology is well proven. Distilled oils are widely used in some countries, but this option has not been widely used throughout the U.K. because of commercial pressures.

In particular, the very low crude and blended oil prices which have existed over the last few years have hampered the rapid expansion of this technology.

Essentially a feedstock is pretreated and then distilled to yield a light fraction, a lube oil fraction and a heavy bituminous residue high in asphaltenes, post treatment is then needed to ensure that a consistent lube oil product of a light colour, low odour and with a pna/pah content equivalent to the original lube oil is obtained.

ECONOMICS

The economics of the three processes arises as follows:

1) Recovery Secondary fuel oil offers a major cost saving of typically 25-40% over the cost of purchasing virgin fuel oil.

2) Recycling Here the economics are more complex. Again cost savings of typically 25-40% against the discounted costs of the new oil are usually achievable. (The discounted new oil price is the purchase price of the oil less its final value (if any) as a secondary fuel oil).

3) Distillation The low cost of crude oil and blended lubes tend to make the recovery of many oils by distillation economically disadvantageous. Niche markets do exist where distillation is viable.

COMPETITIVE ISSUES

1) Recovery Many companies produce secondary fuel oil.

2) Recycling Three companies in the U.K. offer an oil recycling service. Lanstar are the leaders in this technology. Usually the main competition comes from the original new oil supplier who is anxious to protect his market share (lube oil usage in the U.K. has fallen by 20% in the last 2 years).

3) Distillation This is largely an undeveloped market in the U.K. which will only be developed when the customer base is prepared to use a proven quality distilled base oil and market its environmental advantages.

THE FUTURE

My own views are that the use of recovered oil will reduce due to a tightening of controls on stack emissions.

Oil recycling will grow as companies learn to accept their environmental and commercial responsibilities. This will be hastened if they go down an objective scientific route and evaluate the high quality, carefully segregated recycled oil on merit; rather than operate from a position of prejudice which intimates a recovered oil is second hand and this second rate.

Oil distillation will, at some point, boom but only when the commercial factors are favourable and again when the reluctance to use a recovered product has been overcome.

I hope that this presentation has been enjoyable and interesting and thank you for your attention.

Why Recycle?, Rainbow (ed.) © 1994 Balkema, Rotterdam, ISBN 90 5410 367 1

Recycling lead from batteries

J.R.Ainley & H.Rimmington
Britannia Recycling Limited, Wakefield, UK

ABSTRACT: Traditionally up to 100,000 tonnes of lead/acid batteries have been collected and recycled in the UK annually. Within the last two years following the introduction of the Environmental Protection Act, Duty of Care requirements and site licensing considerations, this annual tonnage has been noticeably reduced.

In the wake of new and expensive lead recycling technologies being introduced in the secondary lead industry, any high capital investment plans are becoming very questionable.

As market forces continue to have an adverse input on commodity prices, it is clear that the economical recycling of lead/acid batteries may become a 'thing of the past'. This then becomes a matter for some concern in relation to any potential environmental impact that may result.

1 INTRODUCTION

A vast array of legislative requirements concentrate on the protection of the environment and on wastes in particular and as such, the need to encourage recycling has become more apparent.

Recycling is recognised to be an economical and environmentally acceptable way in which to manage liquid and solid wastes. Many other benefits of recycling are being discovered by everyone, particularly as demands are constantly increasing on domestic and industrial landfill sites. This is most certainly the case where hazardous wastes are concerned.

There are now more and more containers for paper, glass and plastics than ever before and recycling has therefore become an accepted practice.

We do, however, often forget that many metals like lead, copper and aluminium have been successfully collected and recycled for most of this century.

Lead is especially well suited to recycling due to its general accessibility (its traditional uses) and also due to its relatively low melting point and good workability and handleability. Lead, therefore, boasts an exceptionally high recovery rate.

It should, nevertheless, be pointed out that many lead products have little or no chance of ever being recycled, ie underground or underwater cable sheathing and gasoline additives. This loss represents about 6% of the total lead consumed.

The lead recycling industry relies upon lead/acid batteries for its raw materials and therefore high recovery rates are extremely important.

Changes in environmental legislation may not always encourage the recycling of some materials, particularly when those materials are affected by commodity prices and other similar market forces.

2 THE LEAD POSITION

The global consumption of lead is currently some 4 to 4.5 million tonnes per annum (see Figure 1). Out of this total, over 50% is derived from secondary lead processes.

Secondary lead plants around the globe, rely upon materials such as drosses, residues, dewatered sludges and last but certainly not least, lead/acid batteries.

In the UK alone, about 100,00 tonnes of lead/acid batteries are collected and treated. The UK also has one of the highest recovery rates in the world at over 90%.

This traditionally high recovery rate, is now under threat from significant changes in market conditions and an ever increasing burden on industry to comply with legislative requirements. As a consequence to this, only 3 secondary plants now operate in the UK, whereas in the not too distant past 9 smelting plants were in existence.

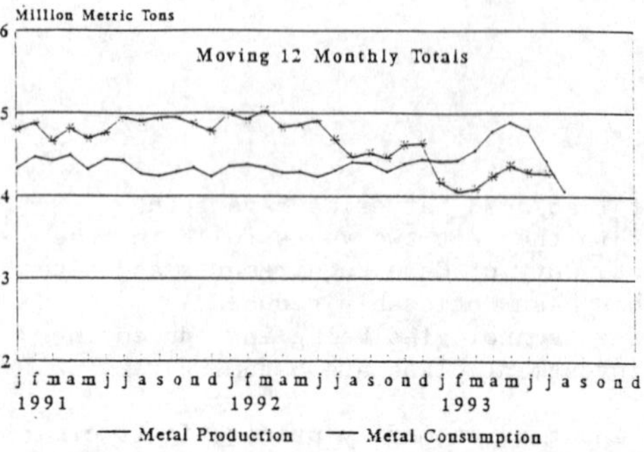

Fig.1 Lead production and demand

Overall levels of lead production tend to fluctuate in line with demand.

It is, however, noticeable that the overall contribution of lead from secondary, as opposed to primary production, has risen by over one million tonnes since 1970, which relates to an annual growth of about 4%. In addition to this, as the availability of lead concentrates may reduce with time, the need for higher recycling rates becomes abundantly clear.

The uses for lead are more traditional than for most metals where some 63% of the total lead consumed is attributed to lead/acid battery manufacture. This is about 4 times as much as the next largest use which is pigments and compounds.

In relation to batteries, a high proportion of the 63% is actually derived from secondary lead sources.

3 SECONDARY RAW MATERIALS AND RECOVERY

The most predominant source for secondary raw materials is lead/acid batteries, therefore recovery rates would be expected to be relatively high. (Table 1) shows how collection efficiencies vary in different countries, however, in the main most are in excess of 90%.

The collection of large industrial batteries is normally arranged by the manufacturer who would usually be replacing old units. This has not necessarily applied to car batteries which have been collected in a very much less organised way.

Traditionally though, spent car batteries have carried a positive value, and this in itself has provided enough motivation and incentive for scrap traders to collect and deliver to the smelters for reprocessing. In recent years, economics have come under review and as such battery manufacturers have become more involved in maintaining high recovery rates.

In many cases the suppliers or outlets for battery sales will accept an old unit in exchange and will return them to a merchant, or alternatively, directly to the smelters.

Table 1. Battery recycling rates in some OECD countries.

	1986	1987	1988	1989	1990	1991
Australia						90
Belgium	77					
Canada						93
France	80			90		
Germany FR	83					>95
Italy	83					
Japan	92.3	92.5	95.2	93.8	92.2	
Spain	83					
Sweden						>100*
United Kingdom	84					93
United States		88.6	91.0	95.3	97.7	

4 BATTERY TREATMENT

Advancements in battery recovery now allow batteries to be broken by mechanical means, which has replaced the traditional axe breaking methods.

There is a choice of breaking machines that are available, but all employ a similar technique of separating and collecting the different component parts of the battery.

The benefit of breaking batteries by mechanical means can be best explained by referring to Figure 3.

Figure 2 clearly demonstrates the 'old' method of battery processing which was greatly disadvantaged, although inexpensive and free from major capital investment.

The introduction of the battery breaking machine opened many new doors for the lead recyclers to explore.

By separating the battery products, selective smelting was possible whereby high yielding materials, ie grid metal and terminals, could be treated to return, not

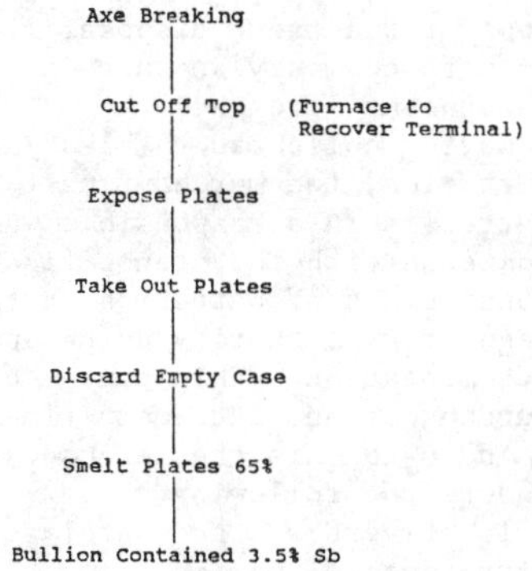

<u>Disadvantages</u>

Large quantities of combustible material present.

Lead lost in cases about 200 g per battery.

Environmentally disadvantaged.

Overriding factor slow - labour intensive.

Fig.2 Battery breaking by axe

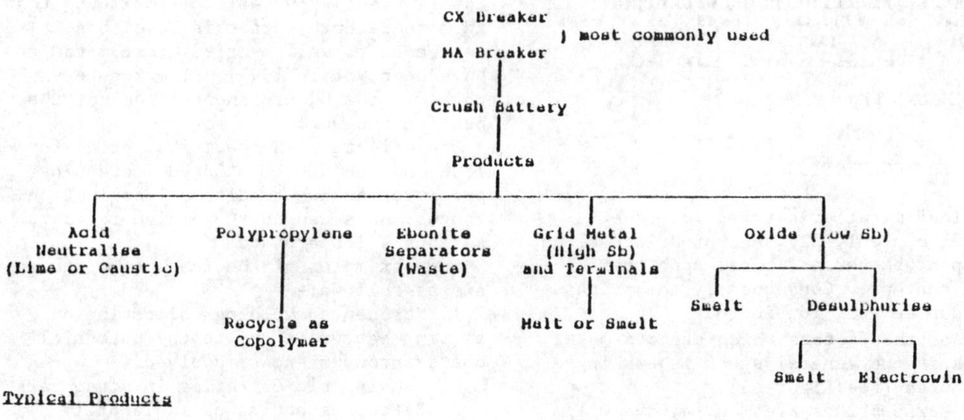

<u>Typical Products</u>

Grids/Terminals	40% of Battery 80% Pb Yield
Oxide	42% of Battery 56% Pb Yield

Fig.3 Battery breaking by mechanical means

only relatively quick furnace bullion, but also the highest proportion of antimony. (Antimony is a common constituent of most battery alloys).

Further options include the recycling of the case material, which is polypropylene. This material can be reprocessed, coloured and reused in a variety of applications.

The most interesting option following battery breaking involves the treatment of battery oxide or paste, ie by pyrometallurgical or electrochemical processing.

Many new smelting techniques have been developed to compete with the more traditional methods of rotary or blast furnace smelting as follows:-

QSL - continuous process
Kivcet - continuous process
Isasmelt - semi-continuous process
Kaldo - batch process

All of these processes have promoted an economical way in which to produce lead metal from lead compounds including concentrates.

There have also been distinct environmental benefits promoted in the operation of these plants over the aforementioned traditional methods of lead smelting.

Various other attempts have been made to perfect, or secure, an economical, non-hazardous route of lead recovery based on electrochemical principles.

Several electrolytes have been used, but relatively few of these have been exploited commercially.

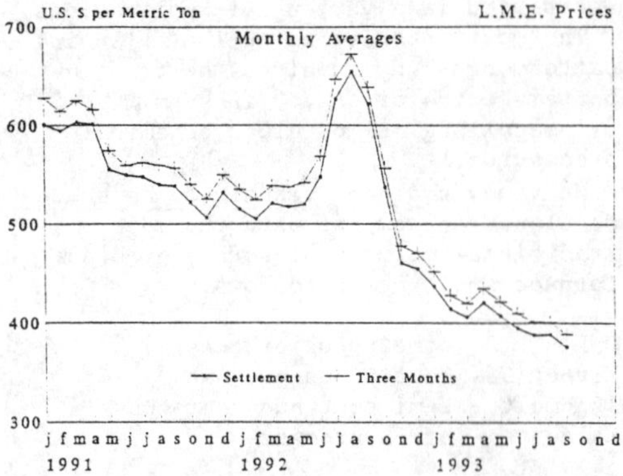

Fig.4 Metal prices

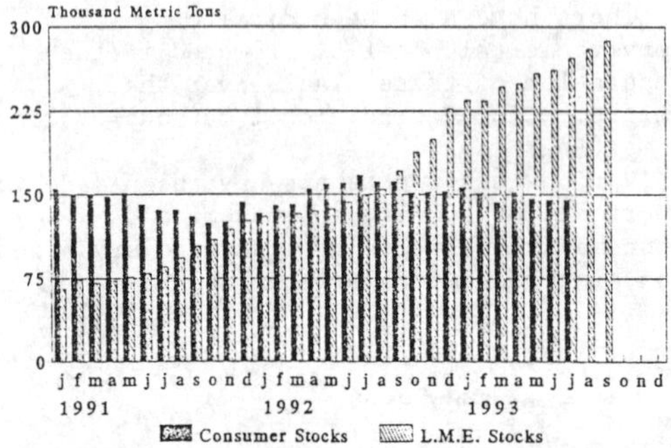

Fig.5 Metal stocks

5 BATTERY COLLECTION

When lead metal prices are low, the smelter can only pay an appropriate amount for spent batteries, based upon their metal content. Consequently the prices paid can be very low indeed.

One of the factors which affects metal prices is the general stock levels in warehouses (see Figure 5).

Low scrap prices are, and always will be, less attractive to those collecting batteries, particularly when legislative requirements place a higher financial burden on collection and storage facilities.

For example, to achieve compliance with the Environmental Protection Act and with the associated 'Duty of Care' requirements, scrap merchants have to provide a storage area which may have to have a surface which is impervious to battery acid and which may need to have a suitable storage or collection container for spillages and/or rainwater. Effluent treatment costs may then be incurred. Added to this, the merchant may have to apply for a waste disposal license or what is more commonly known today as a waste management license.

Having collected the lead/acid batteries, the merchant must transport batteries in a responsible way which might commence with the prenotification of a consignment of batteries to the Waste Regulatory Authority using appropriate documentation. This can be quite burdensome and extremely time consuming.

Added to this the merchant should also ensure the following:-

1. Drivers are adequately trained to carry special wastes.
2. Vehicles carry appropriate plating.
3. Vehicles and containers are sound and leak proof.
4. Documentation is completed and returned.
5. Documentation is retained for up to 2 years.

The general outcome here, is a real and considerable cost consideration to the collector of batteries.

Low commodity prices have a bearing on the destiny of lead/acid batteries and also on the true recovery rates.

Over the last 7 years, the actual recovery rates in the UK have been over 84% (Table 1) however it is believed that in more recent years this level has increased to well over 90% (estimated to have been 93% in 1991). The recovery rates in the UK are amongst some of the best in the world.

The effects of a low market price for lead and the aforementioned increasing pressures on the traditional collection routes, have caused these high rates to be seriously jeopardised.

Some examples of the reasons for this are as follows:-

1. Merchants avoiding collecting or storing batteries due to the potential costs/inconveniences involved.
2. Leaving old batteries in scrap cars.
3. Batteries ending up in landfill sites.
4. Additional batteries being put in to the boots of scrap cars.

The latter point is important, as the market prices for other metals, ie in this case iron scrap from the car, will be more valuable and therefore other metals may be put in deliberately to earn more revenue.

The consequences of these important factors has seen the forecasted lead/acid battery recovery rate fall to about 80% in 1993.

There is, with no shadow of doubt, an unfortunate, but clear, impact on the environment as a result of this.

This impact also features strongly at

the steel making plants where significant and worsening lead emissions would surely be wished to be avoided.

6 SECONDARY SMELTER CONCERNS

There are a number of important issues related to the collection of batteries which the smelters must give every regard to, particularly as these issues may have a significant bearing on the continuation of this traditional industry. The main issues are as follows:-

1. In the absence of high recovery rates and a growing threat of export activity to raw material streams, new investments are almost unthinkable.
2. There are growing threats to the future of the industry while it watches traditional waste stream changes taking place.
3. The growing need to develop a collection scheme in line with the EEC Battery Directive, in the UK, to bring us into line with many other countries. In other words a scheme which will help protect the most vulnerable sections of the lead/acid recycling chain.

7 CONCLUSION

There is a growing concern within the UK that as secondary lead recycling becomes more important in relation to its increasing contribution to the global consumption of lead, that vital components to the business are experiencing adverse changes.

The need to encourage the recycling of lead/acid batteries is of paramount importance, as batteries account for an ever increasing proportion of the metal markets.

Although conventional collection schemes have generally worked well, there is a need to develop new ones or adaptations to existing schemes.

It is hoped that through this development that 100% recycling rates can be achieved, but more importantly, we will help to protect a vital industry and the environment in general.

REFERENCES

International Lead and Zinc Study Group. Monthly Bulletin October 1993.

Wilson, D.N. Journal of Power Sources, 42, P319 to 329 (1993).

Ainley, J.R. Electrochemical Processing, Innovations and Progress Presentation, Glasgow (1993). Unpublished.

Lead Development Association. Various unpublished data.

Why Recycle?, Rainbow (ed.) © 1994 Balkema, Rotterdam, ISBN 90 5410 367 1

Recycle, reclaim, reuse of tyres

Philip J. Howard
Leigh Environmental Ltd, Wolverhampton, UK

ABSTRACT: The paper sets out to examine the role of rubber granulate in the context of tyre reuse. Methods of tyre reuse. Methods of tyre granulate production are described together with their applications for reuse. Other recycle options are discussed and necessary actions described to achieve a higher recycle percentage.

In September 1989, the European Commission delivered a Community Strategy for Waste Management to the Council of Ministers and Parliament. In March 1990 the EC Council of Environment Ministers expressed its approval and backed this up with a Resolution.

In its Resolution of May 1990 the Council welcomed and supported the communication of the European Communities on a Community Strategy for Waste Management. It considered that measures for the prevention of waste at source, the recycling and reuse of waste and the safe and adequate disposal of waste are essential and complementary components of an efficient waste management system.

The Council also considered it desirable, from the point of view of prevention, recycling and reuse, as well as final disposal, to establish action programmes for particular types of waste.

Initially, the Commission decided, as the consequences of a written survey, to consider two priority waste streams. These two streams were:

1. Used tyres
2. Halogenated hydrocarbons

The challenge was to establish programmes that were environmentally acceptable and to include groups that directly effect the production and consumption of that waste stream.

The EEC Used Tyre Project Group held its inaugural meeting in February 1991. Its aim was to achieve a consensus among the various interested bodies concerned and so present to the Council a plan of action that could be welcomed throughout the European Community by producers, consumers, pressure groups and Member States.

The objective of the group was to achieve as near to zero per cent landfill disposal by the year 2000 and to promote prevention and other forms of reuse. This is illustrated in figure 1.

This was demonstrated by reference to the ladder principle.

The bottom rung of the ladder being disposal and

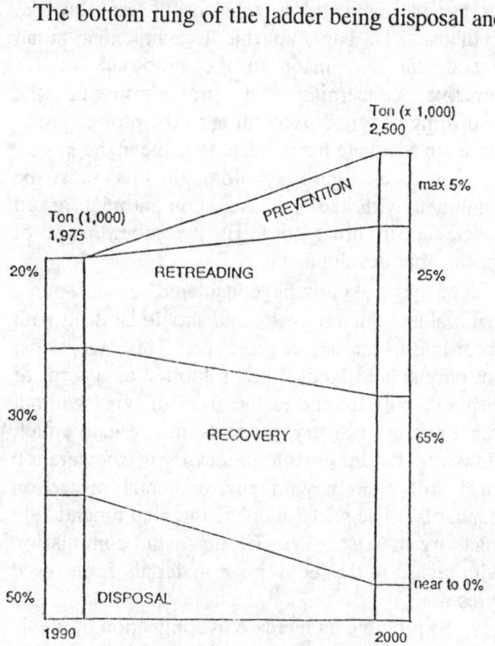

FIGURE 1 – Estimate of the amount of used tyres disposed of in the EC in 1990 and the final target to be achieved.

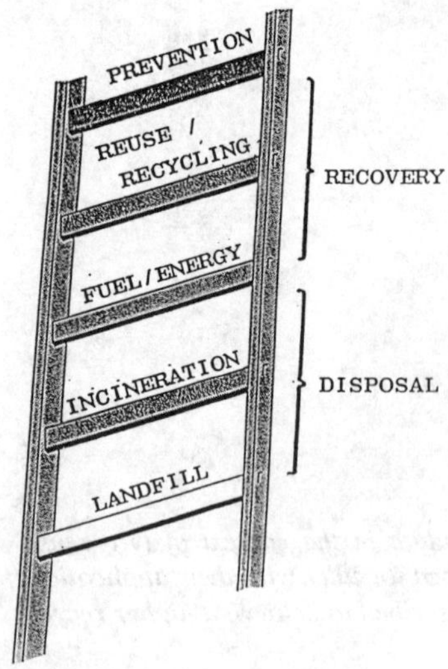

FIGURE 2 – The Ladder Principle

the top rung prevention (figure 2).

The work of the group has resulted in a Commission recommendation and its implementation consists of various measures for prevention, recovery (retreading, reuse recycling and use as a fuel) and incineration or landfill as disposal methods. To back up this recommendation an amendment was made to the proposed landfill directive concerning the prohibition of the landfilling of tyres. According to the project group these are adequate instruments to achieve the targets mentioned in figure 1, although this must be combined with the increased commitment of all concerned in order to fulfil the requirement of 'sustainable development'.

Used tyres should be considered as a source material and not as a waste and should be dealt with according to the ladder principle. This means that not only should landfill be prohibited as a form of disposal, but also the incineration of tyres without either energy recovery or use as an alternative fuel. However, the acceptable processes of incineration must still comply with environmental protection standards. The stated aims of the recommendation made by the Used Tyre Group to the commission with regard to the recovery of materials from used tyres are:

1. To promote an increase in application of whole used tyre and an increase in the useful application of tyres over 140 cm in diameter

2. To promote more application of materials from used tyres (granulate and other materials) by market research, technological research, research on environmental effects and by encouraging fair competition.

3. To promote the incineration of tyres for energy recovery or as a substitute fuel which cannot be otherwise recovered.

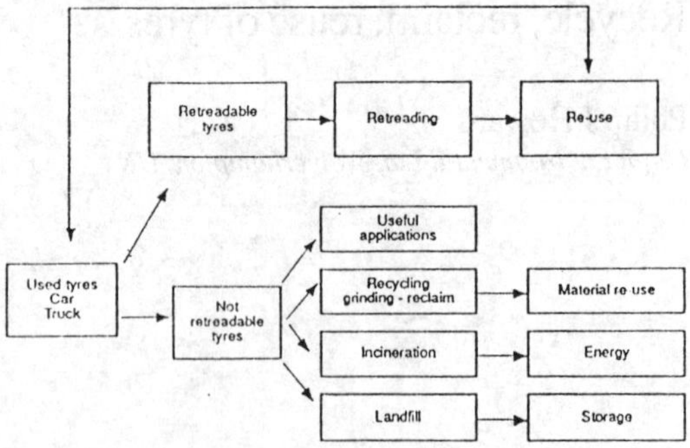

FIGURE 3 – The Life Cycle of a Tyre

Country	tonnes	year
Germany	600,000 *	(1990)
France	362,000	(1989) BLIC
United Kingdom	292,000	(1987) RAPRA
Italy	320,000	(1989) BLIC
Netherlands	65,000	(1986) RIVM
Belgium	70,600	(1989) CLEPIDA
Denmark	26,200	(1986) GENAN
Ireland	19,000	(1990)
Spain	139,000	(1988) SEDICA
Greece	27,000	(1988) SEDICA
Portugal	30,000	(1988) SEDICA
Luxembourg	4,000 estimation	
Total EC	1,955,000	

FIGURE 4 – Breakdown of Production per Member State

The life cycle of a tyre is represented in figure 3.

The volume of tyres generated by the EC at present is 1.95 million tonnes and this is expected to rise to 2.5 million tonnes. The breakdown in production per Member State is given in figure 4.

The amount of used tyres that are either recycled to produce useful materials or used in other beneficial ways is only 14.5 per cent of the total generated volume.

Used tyres can be reused in a large variety of useful ways either in their original form or, after minor processing, in pieces. The following applications are in common use:

1. In civil engineering works (coastal protection, highway structures, erosion barriers).

2. In agriculture to weigh down the covers of silage pits.

3. As boat fenders on harbour walls.

4. For the protection or insulation of pipes and cables.

5. As sound absorbing walls.

6. As insulation in building foundations.

7. As artificial reefs in coastal waters.

In most of these applications the used tyres serve as a substitute for other materials. The preference for the use of tyres in these applications is connected mainly with their mechanical properties, especially their elasticity. Used tyres are all the more attractive because they are of low cost, they cause no environmental problems and can be easily immobilised if responsibly covered with a layer of earth.

Whole tyres and treads can be used to produce granulate or finer crumb material. There are two main types:

1. *Buffed crumb* – this is produced from partly worn (but not scrap) tyres. When a tyre goes for retreading, the rubber is first removed from the tread in form of buffings or raspings. These buffings which would otherwise be waste can be ground and sieved into various particle sizes.

2. *Whole tyre granulate* – this is produced from whole tyres that cannot be retreaded either by grinding at ambient temperature or cryogenically using liquified gas as a cooling medium.

Buffed crumb and the raw material ground from the sole of a tyre form the basis of the reclaim industry. The industry is nearly as old as rubber manufacturing. The consumption of reclaimed rubber in Europe was at its peak in the 1950s, when it amounted to 18 per cent of the total consumption of natural and synthetic rubbers. The tremendous growth in synthetic rubber production with a drop in the price of raw materials brought the production of reclaimed rubber down to a relative consumption figure in 1983 of only 2.5 per cent.

By applying heat and chemical agents to the ground vulcanised waste rubber from tyres the rubber compound can be substantially regenerated into its original plastic state which can then be processed, compounded and re-vulcanised. The process is essentially one of depolymerisation.

The market for rubber reclaim depends on the cost of production and its performance relative to virgin rubber. At the present time the reclaim industry faces many problems, these include:

1. The price of rubber reclaim is too high in comparison with the price of compounds made from new materials.

2. The development of new, cleaner reclaiming processes has not been the subject of significant research and development.

3. The applications for using reclaimed rubber are somewhat limited.

4. The automotive industry is setting increasing quality standards which cannot be met by the manufacture of rubber components using reclaimed rubber. Reclaim is, by definition, of variable composition, since it arises from a variety of different tyres produced by different manufacturers.

5. Only 1 to 2 per cent of reclaimed rubber can be added to new tyres. This is due to the poor mechanical properties of depolymerised rubber.

Whole tyre granulate production is most simply carried out at ambient temperature since it does not require sophisticated technology.

Whole tyres that cannot be used for retreading are first passed through a shredding machine. These machines usually employ contra-rotating shears and have a very high energy consumption.

The pieces of rubber shred exiting from the shredding machines are nominally 50 mm square. They are then passed on to rotating mills which are used in series. By using machines turning at different speeds and using different screen sizes the rubber pieces are reduced in size to a few millimetres. At this stage the material is passed over a magnetic separator whereby the steel wire is removed.

The rubber tyre has now been reduced in size to a nominal 8 mm, but still contains the textile fibre used to support the tread and side wall. To remove this material the rubber pieces are ground down through a series of fixed speed granulators which by varying the screen size reduce the rubber pieces to 3 mm, 4 mm or 5 mm pieces and allows for the separation of the textile using an air classifier.

One such system is described in figure 5. Other techniques involve the use of grinding mills rather than high-speed granulators but the overall result is the same.

The results from this type of operation are highly satisfactory. Yields of rubber from tyres can be expected to be 60 to 70 per cent, depending on the type of tyre being processed. Uses for the steel and textile fibre are limited, particularly the steel, and subject to outside market forces.

Granulate may also be obtained from used tyres by the application of cryogenics. The principle is to take advantage of the properties of rubber at very low temperatures. The system works best for pieces of rubber and thus an initial shredding or de-beading and chopping operation is preferable. The cryogenic medium is usually liquid nitrogen.

The pieces of rubber are then subjected to very low temperatures so that the rubber becomes fragile and the bond between the steel and the textile fibre

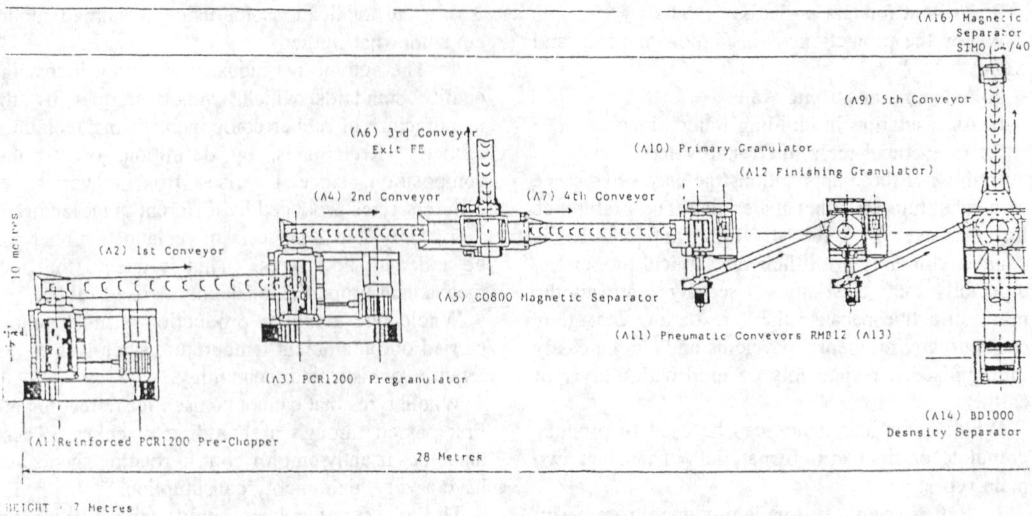

FIGURE 5 – Process Flow Diagram of a Typical Rubber Granulation Plant

is reduced. The frozen pieces are then passed through a hammer mill and screen whereby the rubber is reduced in size, usually 3 mm to 5 mm to allow for separation of the steel and textile fibre using techniques of magnetic separation and air classification. The rubber granules are then passed over a series of sieves facilitating separation into various grades.

Cryogenic grinding has the advantage over ambient grinding in that it does not emit any pollution to the atmosphere. The processes involving ambient grinding require dust filtration to prevent unacceptable releases to the surrounding environment. The main disadvantage with cryogenic grinding is the cost of the liquid nitrogen which makes it uneconomic except for cases where there is a specialist need or there is a cheaper source of liquid nitrogen available.

There are at present companies in Germany, the Netherlands and the United Kingdom involved in operations of this type. The common problem, however, remains the limited market potential where volume is concerned, as only 2 to 3 per cent of rubber available from used tyres is used in granulate applications.

The granulate produced by these methods can be used in the following applications:

1. Additives for asphalt – used as either a binder or filler in asphalt for road surfacing. There is evidence that the addition of rubber powder significantly increases the average life of the road surface. The surfaces acquire greater elasticity, better resistance to cracking and considerable noise reduction.

2. Sports surfaces and playground surfaces – taking advantage of the material's elasticity to relieve ground contact.

3. Solid tyres – used as an inert filler.

4. Civil engineering – bases of roads, particularly in wet areas, or in containment structures. Research into other uses of not only rubber granules but also the byproducts of separation is active at the present time.

5. Friction material for brake systems – used in conjunction with ground glass.

6. Carpet industry – used in the manufacture of carpet underlay.

7. Roofing products – particularly if the granules are further reacted with hot oil.

8. Bound rubber products – used for production of low quality mouldings.

9. Treatment media – research is being carried out to see if granules can be used as a support medium in effluent treatment plants. There is also evidence to indicate that adding ground rubber to hazardous organic wastes can help prevent the leaching of toxic components.

10. Reuse in new tyres – a new type of liquid unsaturated synthetic rubber has been developed which together with a curing agent can be used in concentrations of up to 10 per cent in tyre compounds.

Used tyres can also be used as a fuel. Tyre rubber provides about 32.5 MJ/Kg of heat energy. The most common use for tyres as a fuel in Europe is in cement kilns.

The use of tyres as a fuel in cement kilns has a

number of advantages. All the materials incorporated in the tyre can be completely burnt and absorbed into the cement without any detrimental effect on the quality of the cement. The use of tyres as a supplementary and/or substitute fuel does however require adaptation of the kilns and construction of installations for automatic supply to the kilns. Extensive investments of this nature require a cost recovery period and hence a constant flow of used tyres. This can only be ensured if long term contracts are concluded and cheaper disposal options are not available. The positioning of the cement kilns represent a good geographical spread and thus the cost of transport is contained. The United Kingdom cement industry has not used tyres for a number of years because of economic reasons, but are now re-examining their potential.

Tyres can also be subjected to pyrolysis, or gasification processes designed to produce raw materials and energy. Most recently specialised *waste to energy* plants have been built to burn tyres and provide electricity for use by the National Grid.

In conclusion, used tyres represent a very specific type of waste which is not right to dispose of in landfill sites. Following the principles at the core of priority waste stream management in the European Community then the following policy should be applied to used tyres:

1. As far as possible, they should be retreaded.
2. If tyres are found unsuitable for retreading then reuse applications should be sought either in their existing state or after pretreatment (granulation).
3. Whenever it is not possible to reclaim materials owing to an absence of market, it is then appropriate to recover the energy content by incineration.

Incineration for energy recovery represents a medium term solution of the disposal problem, but in the long term the use of reclaimed materials must be given greater significance.

Why Recycle?, Rainbow (ed.) © 1994 Balkema, Rotterdam, ISBN 90 5410 367 1

Plastics recycling

M.B.Harnett
Breckland Plastics Ltd, Brandon, Suffolk, UK

ABSTRACT: This paper briefly discusses some of the common misconceptions regarding the recycling of plastics and the recyclate itself, both by the general public and by some plastics manufacturers. The viability of recycling plastics scrap from various sources is considered and examples of various kinds, both viable and non-viable are given. The need to promote awareness of the practicalities of plastics recycling is discussed and examples of some misunderstandings are given.

It is suggested that the way forward should be threefold:- cut down on unnecessary use, positively promote the use of recyclate and use controlled incineration with energy recovery for non-viable waste.

What do we mean by recycling? More importantly what do people understand by the term recycling? To the general public, recycling means gathering used items and putting them in a collecting bin. Unfortunately, this limited conception seems to have permeated into some government circles. Recycling is not just collecting waste material. To be successful the material, be it glass, metal, paper or plastic, must be reprocessed and re-used. A market must exist or be created to use the recycled material. This market must be capable of expanding as the supply of recycled stock expands, it must be competitive on price and quality, and must be profitable.

Why do we recycle? For two main reasons, primarily to conserve the planet's resources, and secondly to save energy. Some people, in their eagerness to carry out the primary task, tend to forget or ignore the second part. The subject of this paper is specifically plastics recycling, although many of the remarks also apply to other forms of recycling. I want to briefly cover some common misconceptions regarding plastics recycling, it's viability, the need for general awareness of the problems and possibilities, and the way forward.

One of the commonest misconceptions amongst the general public is that "plastic" is basically the same material which can happily be mixed together. The truth is that there is such a bewildering variety of different plastics today, that even the so-called experts are hard put to distinguish them without, in some cases, very sophisticated equipment.

Many people have only a vague idea of what is involved in recycling, their only contacts with it being collecting bins or a statement on a package that it contains "up to 25% recycled material". They have a vision of a large vessel into which the recyclate, be it metal, glass or plastic, is thrown, with clean material coming out of the bottom. The contamination, (food residues, dirt, paper labels etc.) somehow disappearing in the process.

Among some manufacturers of plastic items there is a feeling, often subconscious, that recycled plastic is second rate material, only good for "rubbish" jobs and that "we never use it as we only have the best". Pressure from environmentalists and the dissemination of information from bodies such as the B.P.F. and the C.B.I. are helping to remove this prejudice but it is still there in quite a few cases.

Turning now to the viability of plastics recycling, I shall make a statement which, however obvious, seems to get overlooked as far as recycling goes. As with any business, to be successful it must be profitable! Subsidies, grants, etc. are fine in the short term to set up the business, buy land, buildings, etc., but once the plant goes into production it has got to be able to run at a profit. We all know what happens to subsidies and grants when hard times come! To achieve a profit the plant needs regular supplies of reasonably clean single materials in bulk. They must be in bulk from any one source or the cost of transport will rule them out. For instance, if it costs £100 to transport 18 tonnes from A to B it's likely to cost at least £60 to transport half a tonne over the same distance. Small amounts of material scattered over wide areas are, unfortunately, not worth collecting commercially, as several specific schemes have found to their cost.

I mentioned single materials just now. This is the key factor, because plastics once mixed are very difficult to un-mix, and whilst certain broad groups can, if necessary to be processed together, most plastics are so diverse in their processing conditions as to render it impossible to mix them. The different types of plastics can be separated by various means but it is either labour intensive and very time consuming, or uses up a lot of energy. Either way the cost of the finished product is likely to rise to the point where it is uneconomic. Cleanliness is important too, since, unlike metal or glass, the temperatures at which plastics are melted are not high enough to burn off or separate the contamination such as food residues, paper labels etc. These must be removed before processing, if a clean, re-usable product is to be made. Whilst it is possible to clean off food residues, oil, shampoo, dirt, paper labels, metal foil, etc. etc., this is also time and energy consuming. The final result may be clean, usable product, but in order to recover the preparation and processing costs, it would have to be sold at a higher price than for virgin raw material. The paper industry discovered this the hard way. When recycled paper cost the consumer more than "new" paper they stopped buying, the waste paper price collapsed and you could not give the stuff away.

Some specific initiatives have been pioneered in the plastics field, such as collecting coffee cups and plastic drinks bottles. Unless the recycling plant is very close to the collection points, the high cost of transport, due to the very low weight to volume ratio, means that the operation will always be non-profit making. While the companies concerned are prepared to support such ventures then all power to them, but as mentioned earlier, in hard times when economies must be made, an operation of this sort would obviously come in for some very hard scrutiny.

Many of the plastic containers used by the retail trade today are made of several different materials, (usually for very good reasons) are often in several colours, printed and decorated, either directly onto the plastic or via glued-on paper labels. A well known brand of shampoo comes in a high density polyethylene bottle with printing, a polypropylene screw-on-top in assorted colours according to the flavour, a low density polyethylene suction tube in natural, a white nylon valve and a metal spring to provide the pump action. To recycle this it would have to be dismantled by hand, the nylon valve and metal spring cut out of the top, and each item then thoroughly washed to remove shampoo residues. The only sensible solution for dirty, used plastic containers of this sort is controlled incineration with energy recovery. Plastic oil bottles, yoghurt, pot noodle or margarine tubs, boil-in-the bag pouches are far better used to provide energy rather than using up a lot of energy, time and money trying to clean and sort them.

Plastics recycling comes into it's own with manufacturers scrap such as film off-cuts, rejects, sprues, redundant stocks etc. This scrap is usually of one material, sometimes of one colour, and rejects and off-cuts often occur before any printing or decoration is added, so it is clean material. It can be picked up, in bulk, from a single point and so is ideal for economic recycling. It is in these sorts of areas that the effort should be concentrated.

What needs to be done to promote the awareness of what is viable in plastics recycling? Principally, those who set standards and draw up specifications need to be shown, that, provided recycled plastics pass all the necessary tests, there is no reason to bar them. Historically it is true that the plastics available thirty or forty years ago, were often not up to standard when reprocessed. Great improvements in quality mean that this is certainly not true today, but unfortunately the attitude of recyclate being no good still persists in some quarters. Much has been done to dispel this attitude, notably by the B.P.F. Recycling Council, but there are still areas where recyclate is only reluctantly accepted often with restrictions such as maximum 10% addition.

There is also a total lack of understanding of the practicalities of plastics recycling among some of the environmental pressure groups, and this needs to be corrected. I saw an article a while ago in which the household of the future was depicted. In the kitchen of the future was a bank of disposal chutes labelled "Plastic" "Metal" "Glass" "Paper". Notice that ALL plastics, ALL metal, ALL glass and ALL paper are just thrown in together. It was envisaged that from each kitchen these chutes would lead down to a common collection area, which would be emptied periodically by some unspecified person. Presumably the writer assumed we would all be living in high rise blocks by then! The article went on to say that these saved items would be recycled, thus saving the Planet's dwindling resources of energy and raw materials. The fact that it may well be necessary to use more energy collecting, cleaning, sorting and processing them, than by using new material seems to have escaped the notice of the writer completely! It is this sort of vague attitude which must be corrected. I am sorry to say that this attitude also seems to extend to the politicians and E.C. bureaucrats. Their main concern is to be seen to be "green". Thus, any scheme which hints at collecting used items is looked on with great favour, even if its totally impracticable and will cost a lot of money. It is no good just collecting assorted used plastics of all kinds, with contamination, paper, metal, labels etc. and expecting someone to do with them. Germany tried this and ended up paying people to take the scrap away. In many cases there is a strong suspicion that it was just dumped in Third World Countries.

The way forward should be through a three-fold push:-

1. Cut down on the unnecessary use of plastics.

2. Positively promote the wider use of recycled plastics whenever possible.

3. Use controlled incineration to recycle the energy locked up in otherwise useless plastic (and other) items.

There are areas where economies of plastics use could apply. Modern plastics are greatly superior even to those available a few years ago. Some items could be redesigned with thinner sections and still retain the desired strength etc. Packaging is usually the section which comes in for most criticism in this context. Where a perfectly ordinary, not very expensive, ball-point pen is encased in a neat plastic box, placed inside a cardboard sleeve, shrink-wrapped and, when sold, put in a separate plastic bag, the cries of overpackaging could be justified. However, as with all things, the danger of going too far must be avoided. Do we want to return to the days when the milkman ladled out your milk from a big churn into your own container? Or the grocer threw the cat off, then scooped up tea leaves from a big sack into a paper bag? I think not. Many of today's comestibles rely upon modern packaging methods to retain quality, freshness and hygiene and we must not jeopardise that.

The sensible way to increase the amount of materials which are recycled to is to actively promote the use of recyclate whenever possible. If more standards and specifications actively encouraged the use of recyclate rather than not actually prohibiting it, and if, in suitable cases, recyclate only was specified, manufacturers would quickly demand more and better methods of collecting more suitable material and recycling would improve and increase through a market-led progression.

With the best will in the world, plastics can only be reprocessed a limited number of times before degrading. All we can do is add antioxidants etc. to prolong their life as much as possible. Eventually the material will be of no further use for reprocessing. It is at this point that the third alternative should be employed.

Controlled incineration, with energy recovery must be regarded as a form of recycling. Surely it is better to recycle the energy locked up in the material rather than throw it in a hole in the ground? Similarly, very dirty items should be recycled by incineration. In this way the dirt, food residues, paper labels, etc., can all contribute to the energy recovery, rather than us having to use up energy to get rid of them. Whilst the amount of energy that contaminants may contribute is very small, the amount of energy required to remove them from the plastic is certainly not small.

To increase the use of incineration however, the tipping cost must be less than that for landfill. At present, for many companies it is cheaper to send unsuitable material to landfill than to send it for incineration.

When burning plastics is mentioned, most people think of columns of thick, black, oily smoke. Modern incineration plants have very sophisticated devices to ensure a clean burn and often the flue gases are as clean or cleaner than the air going in. There is, I feel, great scope here for the dissemination of correct information as to exactly what is involved in a modern incineration unit. If much of the waste which now goes to landfill could be properly incinerated and the energy used to produce electricity, millions of pounds could be saved and large areas of land not subjected to the aggravation caused by rubbish dumps.

Why Recycle?, Rainbow (ed.) © 1994 Balkema, Rotterdam, ISBN 90 5410 367 1

The scrap rubber problem

Kevin P. Jones & Kristina Lawson
Rubber Consultants, Hertford, UK

ABSTRACT: It is possible to recycle many materials, such as glass, back to their original states with no loss of properties. Others, such as paper and many thermoplastics, can be recycled into something with poorer, but still useful, properties. Rubber is not nearly as amenable to recycling: rubber reclaiming requires high energy and chemical inputs which are difficult to justify within the current economic framework. Defering the recycling problem is an important contribution. This can be achieved either through product life extension or through renovation, as in the case of tyre retreading, The latter is important in the aircraft and road haulage industries, but is still declining in the passenger car sector. Controlled combustion probably forms the best means of ultimate disposal, although rubberized bitumen in roads could be an alternative if government intervention is provided.

1 INTRODUCTION

Rubber is not a simple material to recycle and tyre dumps are recognized as being a major environmental danger. Fires in tyre dumps are difficult to extinguish and may burn for months. The heavy smoke is not the only hazard as water supplies may also be contaminated. The cost of such fires is considerable and there is mounting pressure not only to discontinue tyre dumping, but to eliminate existing dumps.

1.1 *The composition of rubber products*

A few rubber products (notably medical gloves, balloons, teats and condoms) consist predominantly of rubber which has been modified with a relatively small quantity of chemicals to vulcanize, and possibly protect and colour them. Sulphur is the normal basis for vulcanization, but other materials may be used and vulcanization is normally assisted by the presence of small quantities of other chemicals known as accelerators. The sulphur (and other chemicals) modify the rubber in what is essentially an irreversible process. This is the core problem of recycling rubber. Nevertheless, if the economics are correct then it is possible to reclaim rubber back to a material which is capable of being incorporated within new rubber products.

The majority of rubber products are complex in nature and may incorporate more than one type of rubber either as a blend, or as separate sub-components (as in the case of treads and sidewalls for tyres), or as both. Even products which superficially appear to consist of a single compound, such as extruded automotive weatherstrip, are frequently based on more than one compound to provide a number of distinct properties.

A very small amount of rubber is used in its virgin state. The commonest examples are adhesives based on latex, such as *Copydex* sold in the United Kingdom, and crepe rubber soling: the latter is raw natural rubber. Some highly abrasion resistant linings, as used in sand and gravel extraction, consist of very lightly vulcanized natural rubber. Self-curing silicone rubbers, and some other silicone rubber products are also relatively unadulterated with other materials.

The bulk of rubber products contain not only chemically bound crosslinks introduced through vulcanization, but also a range of other materials to protect, soften, strengthen, and/or cheapen them. Protective agents are designed to lessen the effects of heat, light and ozone. These tend to be high value materials and are present in small quantities. Softening is achieved by the use of mineral oils: in most cases only small quantities are present, but some automotive sound-deadening panels contain very high quantities of oil. Internal reinforcement is typically achieved with carbon black, although silica may sometimes be used. Cheapening is achieved by inorganic fillers based on china clay and whiting: large quantities of such fillers are used in carpet backings. Flame resistant elastomers tend to incorporate antimony trioxide. Radiation resistant rubbers incorporate high levels of lead.

Both textiles and metals may be used to reinforce the rubber: indeed, one of the earliest rubber products, the macintosh, was based on a sandwich of rubber solution between cotton. Such reinforcement may be relatively discrete as in the case of some bearings which consist of alternate layers of rubber and metal (or glass fibre) or may be intimately embedded within the rubber as in the case of the reinforcement in tyres and belting. Anyone who has experienced a tyre blow out will have witnessed the twisted mass of rubber and metal. With both types of reinforcement the designer has to ensure that the rubber and its reinforcement are as inseparable as possible. Under destructive testing many rubber-to-metal bonded components fail not at the bond, but within the bulk of the rubber.

Even apparently simple products like hot water bottles may contain a rubber or plastic insert: hot water bottles require a filler. Thus even if rubber *per se* was simple to recycle (which it is not) there would always be the problem of removing the other materials. Sometimes it is more economic to burn the rubber and to recycle the associated metal.

1.2 *The range of elastomers*

The commonest elastomers, in approximately ranked order, are natural rubber, styrene-butadiene rubber (SBR), polybutadiene, synthetic polyisoprene, butyl rubber, EPDM, nitrile rubber and polychloroprene. With the exception of natural rubber all are based on petrochemicals. Synthetic polyisoprene is now produced in large tonnages only within the CIS: its properties and uses are very similar to those of natural rubber. The first five materials form the major elastomeric ingredients of tyres: in the case of polybutadiene and butyl rubber tyres form virtually the sole outlets. EPDM is widely used in non-tyre automotive applications and also forms the basis for a substantial quantity of thermoplastic elastomer. Nitrile rubber is the commonest form of oil resistant rubber, but it has been displaced from some of the more demanding applications by a range of specialist, expensive elastomers. Polychloroprene is often known by its Du Pont tradename of *Neoprene*: it is used where ozone or flame resistance are required. The approximate annual consumption statistics for the more common types of elastomer are shown in Table 1.

Table 1. Annual consumption of major elastomers

Elastomer	Annual consumption (million tonnes)
Natural rubber	5.630
Styrene-butadiene rubber	4.545
Polybutadiene	1.370
Cis polyisoprene (synthetic)	0.75
Butyl rubber	0.55
EPDM	0.600
Nitrile rubber	0.350
Polychloroprene	0.285

It should be noted that the figure for styrene-butadiene rubber includes 331 000 tonnes of SBR latex and 1 278 000 tonnes of carboxylated SBR latex. These latices are used in coated papers and in the backings for tufted carpeting, where they are admixed with high amounts of calcium carbonate filler. About 10% of natural rubber is also consumed in the form of latex: one of the major applications is dipped goods, such as gloves, catheters and condoms. There is a wide range of high value, high performance speciality elastomers. These include chlorosulphonated polyethylene, often known by its Du Pont tradename of *Hypalon*, silicone rubbers and fluoroelastomers. Silicone rubbers are notable for their ability to withstand higher temperatures than most elastomers; they are also widely used as self-curing materials in the form of sealants (a form which they share with the less expensive polysulphide elastomers). Fluoroelastomers are finding increasing employment within automotive seals and hose as they are resistant to hydrocarbon oils at high temperatures. Most of these speciality elastomers are produced in quantities of less than 100 000 tonnes per annum. By the nature of their applications (small components within severe environments) most are not likely to be easy to recycle.

Thermoplastic elastomers are growing in importance. The current global market is in excess of 800 000 tonnes. They are tending both to displace some traditional elastomers and some thermoplastics. Several types are available: one group is based upon butadiene and styrene, another upon polyurethane, and another upon EPDM or natural rubber. Thus as a mass they chemically resemble thermosetting elastomers, but they do have a much higher potential for recycling as they are capable of being reformed under the action of heat. Thus, it is relatively simple to recycle scrap produced during manufacture. Scrap components, such as automotive weatherstrip, may have to be cleaned before they can be recycled, however. One advantage which they possess in terms of combustion is that they contain little, if any, sulphur.

Polyurethanes sometimes fulfil an elastomeric function, but substantial quantities are employed in essentially non-elastomeric roles, such as cavity wall insulation. Foamed polyurethane has tended to displace foam rubber based upon natural rubber and SBR latices from most upholstery markets except mattresses and pillows. Polyurethane thread, Spandex, competes with thread based upon natural rubber latex for many applications in garments. Polyurethane is used in soling for footwear and for some industrial tyres. The highly toxic gas, hydrogen cyanide, is produced during the combustion of polyurethanes.

From the point of view of the recycler presented with a pile of rubber then the most probable overall composition is likely to be styrene-butadiene rubber (around 75% of which is based upon butadiene) and natural rubber. The two polymers blend easily and there is no difficulty in using the polymers in conjunction with each other. Both fulfil specific roles within tyres. For instance, SBR is used within the tread to achieve wet grip: natural rubber is used within winter tyre treads to enhance grip on ice. As tyres form at least 60% of discarded rubber products then the typical tyre dump consists of polymers based upon butadiene or natural rubber, but it must be remembered that other materials are also present in considerable quantities.

1.3 *Tyres*

Tyres vary greatly in weight: passenger car tyres weigh about 7kg; light truck tyres about 24kg; heavy truck and bus tyres about 57kg and some very large earthmover tyres up to one tonne. Clearly the passenger tyres are the most common and the giant earthmover tyres the least. Some statistics for the annual consumption (in 1992) of passenger and truck tyres are given in Table 2.

On a global basis, about 15 million tonnes of tyres are produced per annum and this represents about 9 million tonnes of rubber, 1 million tonnes of steel tyre cord and 3 million tonnes of carbon black. Depending upon the tyre type, the steel content of a tyre varies between 5 and 15%: truck tyres tend to contain more steel.

Care needs to be taken in the interpretation of tyre statistics. For instance, Japanese vehicles are much smaller than those found elsewhere. Furthermore, because of the shortage of roads most Japanese vehicles are used far less than those in other developed countries. Road conditions differ markedly between Europe and the USA and this not only leads to lower tyre life in Europe, but also to different forms of tyre construction and composition. Wet grip is far more important in Europe.

2 SOLUTIONS

2.1 *Product life extension*

The automotive industry, which is the major customer for rubber products, is seeking to eliminate the replacement of components during vehicle life. This has reduced the number of elastomeric components which require to be replaced, or reduced the frequency with which they are changed. Windscreen wiper blades, exhaust hangers and tyres are the most frequently changed items, although in the USA tyre life for passenger vehicles is being extended to that approaching the vehicles themselves.

Product life extension is being achieved by the better design of elastomeric products. For instance, the application of finite element analysis has enabled designers to avoid stress concentrations within components which are a potential source of failure. Elastomers are more closely specified and better formulated, and there has been a tendency to use more specialized elastomers to give greater product performance, especially within the engine compartment.

The repair of rubber products is relatively uncommon. Punctures within tyres may be plugged. Conveyor belts, hovercraft skirts, rubber floors and the rubber linings used both to protect against the action of chemicals or abrasion may be patched. The bulk of rubber soled footwear, such as trainers, is not repaired: the major exceptions are high quality shoes and high quality waterproof boots. By far the most important form of product life extension cum recycling is tyre retreading. In this process the old tread is buffed off the carcass and a new tread is applied.

2.2 *Retreading*

The bulk of commercial aircraft landings and take-offs are performed on retreaded tyres. Moreover, aircraft tyres are re-capped (this specialist industry prefers this term to *retreaded*) several times. Eight times is a typical figure for a passenger airliner. Admittedly, the surfaces upon which such vehicles land are monitored carefully and the amount of turning is limited to low speed manoeuvres on taxiways,

Table 2. 1992 Consumption of tyres (millions)

Country	Passenger car	Truck & bus
Germany	55.048	6.549
France	32.547	3.545
Italy	25.411	3.384
United Kingdom	35.547	2.071
Benelux	13.420	1.515
USA	211.165	39.987
Japan	92.498	37.263

Table 3. Markets for retreaded tyres (millions)

Country	Year	Commercial vehicles	Passenger cars
France	1991	1.1	
Germany	1991	0.83	2.4
Italy	1989	1.16	3.5
United Kingdom	1991	1.0	4.5
USA	1991	22.3	8.4

but the speeds attained and the tyre surface temperatures on landing are high. The reason for retreading is that wear rates are very high as is clearly shown by the black streaks on most runways and aircraft tyres are expensive. A reconditioned aircraft tyre costs about one third of a new one.

Earthmover tyres are also very expensive and the tyres are designed so that the lugs can be replaced or the tyres can be retreaded. In some countries, including the United Kingdom, it is possible to re-groove truck tyres, although few countries permit the re-grooving of passenger tyres. It is relatively normal for truck tyres to be retreaded, although practice varies from country to country. US truckers employ more retreads than their European counterparts, probably because of the greater percentage of long-haul work: local work tends to involve damage to the tyre carcass and sidewall through scuffing.

Retreading is an excellent way of extending product life and deferring the problem of disposal. Even the major by-product of tyre retreading, the buffings from the former tread, finds a ready market in outlets such as the surfaces for children's play areas as the buffings do not contain metal which might cause injury. Other uses include sports surfaces, traffic cones, and as a component within carpet underlay. Specialized sporting mountaineering boots enjoy the extra grip obtainable from incorporating the buffings from re-capped aircraft tyres.

In virtually all major European countries and the USA the market for passenger car retreads is still declining. In 1986 remoulds accounted for 21% of German passenger tyre replacement sales: this had fallen to 9% in 1991. In Italy the production of passenger retreads fell from 7.4 million in 1980 to 3.5 million in 1989. One major reason is the ready availability of low cost, imported replacement tyres in most of the major markets. Part of the reason is one of image. Part is one of speed restrictions: most retreads are speed limited, but these limits tend to coincide with national limits.

Most passenger tyres are designed without consideration of the need for retreading: this is in marked contrast to truck and aircraft tyres. A very high percentage of passenger tyres are driven to a point beyond which they are retreadable. Many passenger tyres in the United Kingdom are being used in an illegal state: the Tyre Industry Council estimates that there are twelve million illegal (and dangerous) tyres being used in the UK. The design constraints are largely due to the automobile industry's quest for lower overall weight, hence lower fuel consumption. Retreadable passenger tyres would need to be heavier and would probably lead to higher fuel consumption.

The aircraft, commercial vehicle and earthmover retreading markets are assisted by the willingness of the vehicle operators and tyre manufacturers to cooperate in the retreading process. Within this sector it is not unusual for major tyre manufacturers to participate in the retreading process. For instance, Michelin is a major servicer of aircraft tyres. Goodyear in the USA has a major involvement in truck tyre retreading. Truck tyre servicing is a major business and Bandag (a major franchiser of retreading equipment) cooperates with its customers to ensure rapid and simple replacement of tyres on high value vehicles. Major trucking organizations normally employ a tyre service company to optimize the use of their vehicle fleets. These factors are absent from the private car sector.

The routes towards increasing the market for passenger tyre retreads would appear to lie with government or inter-government legislation. The imposition of tougher tread depth limits would both reduce accidents from skidding, especially on wet roads, and would both ensure a larger supply of retreadable tyres. The changed economic prospects might encourage the tyre industry to produce passenger tyres with more substantial carcasses, more suitable for retreading. In effect the motorist would have to pay marginally more, but with the benefits of greater safety and a reduction in the tyre disposal problem. The US Environmental Protection Agency wants to see as much retreading as possible as retreads score high in terms of recycling: more than 75% of the retreaded tyre represents recycled materials.

2.3 *Further uses in original form*

Unfortunately, the re-use of rubber in its original product form is limited. Old conveyor belting can be used to line river banks to inhibit erosion. Old tyres can be used to assist in the formation of reefs or to act as artificial reefs for fish farming. Old tyres can be used intact or cut up to form fenders either on the side of jetties or on boats. They can be used to act as anti-noise barriers along the side of motorways. They can also be cut up and used as crude forms of footwear in very poor countries. To an extent, life expired tyres from motor vehicles may find further use on animal-drawn trucks in poor countries. With the exception of the reefs, most of these 'solutions' merely defer the disposal problem.

2.4 *Crumb from scrap rubber products*

There has been interest in the use of rubber to modify the properties of bitumen both for road service dressings and in roofing for over fifty years. Most rubbers have been considered including expensive materials, like polychloroprene and the newer thermoplastic rubbers. Elastomer modified bitumen can be more wear resistant, more skid resistant and can assist in reducing fatting up at high temperatures and brittle fracture at low temperatures.

Vulcanized scrap in the form of crumb can be used instead of virgin rubber. It has the advantage of being far cheaper – one of the factors which has inhibited the uptake of rubberized bitumen has been that a relatively expensive material (rubber) has to be mixed into a low cost material (bitumen). The construction and maintenance of roads tends to be organized by central and local government. Although the problem of tyre dumping is normally monitored by a different sector of such organizations, the fact that both fall within their remit should assist with the uptake of scrap rubber within road surface dressings.

The United States Congress introduced an Intermodal Surface Transportation Efficiency Act which may eventually stimulate the use of recycled rubber in road surface dressings: in 1994 each US state would have been required to incorporate at least 5% of recycled rubber in bitumen. The amount would have risen to 10% in 1995, 15% in 1996 and 20% in 1997. It is hoped that this would have eliminated tyre dumping. As is common with such US legislation, the administration has recently postponed its implementation as 'The costs of rubber pavement compared to conventional asphalt are concerns that have been raised by many state transportation departments'.

3 RUBBER RECYCLING

All forms of rubber recycling have to face the problem that some fifteen million tonnes of elastomer, or thirty million tonnes of rubber compound, are transformed per annum on a global basis. This is a trivial quantity compared with the vast quantities of coal, natural gas and oil which are consumed to produce electricity. The topology of tyres makes them exceedingly difficult to store or to use as landfill, and this same characteristic makes them difficult to transport in a used condition or to otherwise handle them in a recycling operation. Shredding makes the processes far simpler, but is energy intensive, especially if cryogenic techniques are employed. Any rubber recycling operation will always have to be based upon the concentration of scrap and transport costs will form a considerable element within the overall economics.

A further problem is that raw rubber prices are relatively low. Some indication of the effect of low price may be gained from the *decline* in the production of natural rubber in Malaysia: output has fallen from a peak of 1 661 600 tonnes in 1988 to 1 217 500 tonnes in 1992. The real cost of raw rubber, both natural and synthetic, has fallen over the past decade. The export values of SBR from the USA actually fell from $1008/tonne in 1982 to $813/tonne in 1986, and only recovered to $949 in 1992. Natural rubber prices are slightly more volatile, but a London price of £709.8/tonne in 1988 may be contrasted with £545.8/tonne in 1992. Any recycled product is bound to have a low and uncertain value in such circumstances. Furthermore, if as with the case of reclaimed rubber, it has to compete with the virgin polymer then its economic prospects are dim in open markets.

3.1 *Reclaimed rubber*

Vulcanization of rubber with sulphur was invented by Charles Goodyear in 1839. In the late nineteenth century Arthur Marks developed the alkali digester process in which ground rubber is cooked in a dilute caustic soda solution for about twenty hours: this system removes the textile fibre, much of the residual sulphur, and devulcanizes the material in one step to produce a re-usable material. This technique is suitable for natural rubber-based products, but the process has to be modified by using calcium or zinc chlorides to cope with SBR.

The pan process is suitable for fabric-free scrap, and is thus frequently employed to recycle factory scrap. In this process the ground rubber is mixed with reclaiming agents and then heated in high pressure steam for 5 to 12 hours.

Both the digester and pan techniques are batch processes, but continuous processes have been developed in which the scrap crumb is fed into a screw extruder. Oil and chemicals are added and the process of extrusion generates sufficient heat to devulcanize the rubber.

All of these processes have to be preceded by a cracking stage in which the rubber products are broken down into crumbs small enough for any embedded metal to be extracted with strong magnets. It must be stressed that the resultant product is far removed from virgin rubber, although it may be incorporated into it as a processing aid or extender. Only some of the sulphur crosslinks are broken down during the reclaiming process. On the other hand, a considerable amount of main-chain scission also occurs, and this lowers the molecular weight. Thus the addition of reclaim tends to lower the overall tensile strength, tear strength, resilience and abrasion resistance of products into which it is incorporated.

The reclaiming processes are relatively costly in terms of manpower, chemicals and heat input. Thus the commercial production of reclaim has declined to negligible amounts in most open economies, but reclaim production continues in India and appeared to flourish in Eastern Europe and within the CIS. It is estimated that about 59 000 tonnes of reclaimed rubber are consumed by the Indian rubber industry, as compared with 524 000 tonnes of virgin natural and synthetic rubber. Table 4 shows that Indian production of reclaim increased during the decade 1982-1991.

Table 4. Indian production of reclaimed rubber (tonnes)

Year	Quantity
1982	29282
1983	29815
1984	33480
1985	38550
1986	37575
1987	40835
1988	40175
1989	44125
1990	45180
1991	50120

More recently there has been a revival of interest in reclaiming in Malaysia. This is to cope with recycling dipped goods, mainly gloves, which have failed to meet the stringent international regulations for such products.

3.2 *Modified crumb rubber*

There has been some interest in the modification of scrap crumb by various chemical treatments. Reactive gases (a mixture of fluorine, oxygen, sulphur dioxide) are passed through the crumbs. The modified crumbs are claimed to afford greater adhesion to other continuous phase systems, such as virgin rubber.

Experience at Rubber Consultants has shown that the economics of rubber modification are difficult, even if the modified rubber enjoys considerably enhanced physical properties. Such modifications have to be based on cheap materials and simple processes, and the enhancement in properties has to be worthwhile. Perhaps the most successful elastomer modification in recent years has been the hydrogenation of nitrile rubber: this produces a material which enjoys a much higher operating temperature, together with improvements to other properties. This has enabled the new material to find new markets and a price which covers the costs of modification.

3.3 *Rubber pyrolysis*

There is a large literature on the pyrolysis of rubber products to produce a variety of what are claimed to be useful materials, the organic elements of which are usually claimed to be either oils or potential feedstocks. There is insufficient financial elasticity to employ either expensive chemicals or considerable amounts of external energy to recycle rubber. Ideally any such energy will be provided from the rubber scrap.

4 COMBUSTION

Combustion of rubber products, preferably without recourse to prior crumbing, is probably the most efficient method of handling scrap rubber products. Approximately one third of such products are based upon natural rubber and such combustion will not affect the overall global carbon dioxide budget: the rubber trees are more than capable of re-absorbing the carbon dioxide so produced. Vulcanization is based upon sulphur, and any combustion process will require the combustion products to be scrubbed. For this reason there is interest in the non-sulphur crosslinking of medical gloves by gamma radiation to avoid corrosion of medical incinerators.

The major ingredients of the rubber compound (the elastomer, the carbon black and the oil) are combustible. Between 70 and 75% of tyres is in the form of carbon. Between 5 and 15% is in the form of steel, which should be reclaimable after combustion. Between 1.5 and 2.5% is in the form of zinc oxide which should also be reclaimable. Between 1 and 1.5% is sulphur, which may cause problems, but which is less than the sulphur content of the oil emulsions being imported from Venezuela to burn in British power stations. The gross calorific value of rubber when burnt in oxygen is at least 27.5 MJ/kg. This compares somewhat unfavourably with steam coal (36 MJ/kg), but very favourably with peat (16 MJ/kg): peat it should be noted is used as a source of heat for power generation in suitable circumstances.

A small quantity of scrap tyres is consumed in

association with other fuels in the manufacture of Portland cement, but this takes place mainly on a small-scale and in some countries, notably Germany and Japan. A small amount of scrap is also used within rubber processing factories to provide heat for space heating. The greatest interest lies in the employment of rubber as a fuel to eliminate the scrap tyre problem. Unfortunately, it appears that the amount of heat generated is only sufficient to act either as a source of centralized space heating, or as a minor contributor to electricity generation. Moreover, there is considerable environmental pressure to restrict such operations, either through fear of pollution or from the extra traffic generated in conveying the scrap. It would seem that political pressure, either through taxes or subsidies, will be required to substitute controlled tyre burning for tyre dumping.

Elm Energy expected to start commissioning trials of a 25MW power station in Wolverhampton, fuelled with scrap tyres, in October 1993.

5 CONCLUSIONS

To a large extent national practices on scrap rubber recycling are based on the availability of landfill sites. If these are available, as in the USA, there will still be a temptation to dump. In countries with a shortage of land, like the Low Countries and Japan, there is a much greater incentive to find solutions to the problem. In Japan over 150 000 tonnes of discarded tyres were burned in cement kilns in 1991. As fuels are expensive in Japan there is a spur to consider rubber as a fuel, and scrap tyres (nearly 70 000 tonnes) are even used in metal refining.

Retreading is a highly significant way of recycling rubber and is commonplace in the commercial vehicle and aircraft transportation industries, but the amount of retreading within the passenger car tyre sector is still declining in the major markets. In terms of quantity, passenger tyres form the bulk of the tyre disposal problem. Unless whole vehicle life tyres can be produced, and this appears to be unlikely in Western Europe within the short term, the reintroduction of passenger tyre retreading on a large scale would appear to be the most effective way of reducing the scrap tyre problem.

The increasing adoption of thermoplastic elastomers in non-tyre components should lead to some amelioration of the problem. Ultimately, controlled combustion is the simplest form of disposal and the relatively high amount of natural rubber within the overall scrap budget should ensure that global damage through CO_2 emissions is minimized. The use of scrap rubber in roads is an interesting alternative, but the economics will always be difficult unless, and this is unlikely, there is a shortage of bitumen.

Why Recycle?, Rainbow (ed.) © 1994 Balkema, Rotterdam, ISBN 90 5410 367 1

Challenges and opportunities in steel can recycling

P.A. Neenan
AMG Resources Limited, Birmingham, UK

ABSTRACT: Tinplate is the most common form of material used for packaging of preserved and processed foodstuffs and drinks in the world. Although industrial tinplate scrap has been recycled on a large scale since the turn of the century, post-consumer scrap has remained relatively untouched for numerous technical and commercial reasons and represents the largest single untapped source of ferrous scrap in the world. AMG Resources Limited, the world's largest and oldest detinner, have developed the post-extraction technology combining preparation and detinning after many years of research. The process has been fully-tested and has proven to be commercially and technically sound in both the UK and USA.

1 INTRODUCTION

One of the last single and, until recently, largely untapped sources of ferrous scrap that remains to be recycled effectively is used steel cans. The vast majority of empty cans have found their way into the refuse stream eventually ending up in landfill.

Essentially, the capital and operating costs of extraction, transport, preparation and detinning have inhibited the economic feasibility of large scale recovery. Further, viable solutions to technical problems associated with total recycling have eluded the best endeavours of long-established detinning companies despite considerable research.

Two new opportunities are now available, however, both of which provide a demand led economic recycling route for the used steel can.

Where refuse is processed via an incinerator and it is not possible to extract metals prior to incineration, the ferrous content can be removed magnetically from the residual ash and, as a result of significant development work by the international steel industry, this material will now find a ready market in steelmaking.

The second and preferred route utilizes technology developed by AMG Resources Limited which embraces magnetic extraction from refuse stream processing and detinning to produce the finest quality L.R. steel scrap and ingot tin for re-use by industry. The aluminium content of bi-metallic cans is also recovered and sold for further refining. The first plant at Hartlepool in the UK has been operational for 7 years and has proved to be economically and technically sound and provides the potential for a satisfactory response to both legislation and public pressure aimed at increasing the levels of recycling and conservation.

At this juncture, it may be appropriate to establish a true definition for recycling and conservation. Recycling, whilst a laudible aim, can only be justified in conservation terms if it consumes less resources than would be the case if the recycled product were discarded.

Any commodity can be recycled provided enough money is thrown at it but conservation is, on the other hand, recycling that can be justified in economic and environmental terms.

There is a danger in the current climate of giving credence to collection schemes accumulating material for which there is no market. Many such schemes cannot be justified in economic terms and will result in disappointment and the public participants in such schemes will become disillusioned. The market for recovered metal packaging is mature and is eager to accept properly processed post consumer material. The market is not therefore a constraint in the recycling of metal packaging.

2 HISTORICAL PERSPECTIVE: THE U.K. EXPERIENCE

The concept of recycling metals and particularly cans from household refuse is not new and was in evidence as early as the second decade of this century. Additional emphasis on the strategic importance of recycling used cans came during the war years and commercial recovery developed rapidly during the following twenty years - the material principally being used directly as a furnace feed for the manufacture of iron. This latter point is of particular significance.

Whilst high tin levels could be tolerated by the grey iron making foundry industry in, for example, the manufacture of castings (where the tin was utilised as a pearlite stabiliser); the tin content, which could be as high as 1% by total weight of an individual can, rendered the un-detinned can unsuitable for steelmaking where the presence of tin presented considerable metallurgical problems. It should be noted that the increasing use of bi-metallic cans in the 1970's and 1980's added a further complication in that, whilst certain levels of aluminium contamination could be tolerated in steelmaking, its presence caused considerable problems in iron manufacture which could only be overcome by dilution. Therefore, the U.K. market for refuse derived cans relied on the iron makers. The local authorities recovering those cans could adopt relatively unsophisticated basic extraction techniques. The material recovered was heavily contaminated (30% non-metallics was not uncommon) and was acceptable only because this contamination could be removed simply and cheaply in the Hot Blast Cupola and the tin content tolerated. The volume of metal recovered by this method reached its peak in the mid-1960's when in excess of 100,000 tonnes per annum of cans were being consumed in foundry pig iron manufacture.

During the late 1960's and 1970's, however, considerable change was taking place in the iron manufacturing industry. A severe long term reduction in demand for grey iron castings, together with changes to S.G. and malleable caused the foundry industry in the UK to reduce to one tenth of its former size, with the consequence that the market for synthetic pig iron was insufficient to support the pig iron manufacturers who were forced to close.

The contraction of the foundry industry thus removed at one stroke the only significant commercial outlet for used cans, and the market for this type of recovered metal became extremely limited, with the result that many local authorities in the U.K. ceased to use their magnetic extraction plants.

The only realistic market remaining for recycled refuse derived cans on an industrial scale is the steel making industry and if that market ambition is to be realised the tin content will have to be removed by detinning techniques.

From the 1920's to the 1960's, A.M.G.'s predecessor company (Batchelor Robinson) purchased limited quantities of used cans for detinning to supplement its industrial tinplate purchases. The detinning techniques used were only partially successful in the removal of tin but a ready and stable market was found in the overseas copper precipitation and certain foundry outlets. As these traditional outlets contracted either through changes in technology or the increasing availability of other ferrous scrap sources, attention was focused on prime industrial tinplate scrap the arising of which were increasing and more than sufficient to soak up the existing detinning capacity. The practice of detinning used cans was discontinued.

Indeed, the used can presents the detinner with complex technical hurdles, not least the physical problem of detinning a can as opposed to flat surfaced tinplate scrap but also the many contaminants associated with cans extracted from refuse.

Until the current plant was opened in 1986, no detinner was able to process refuse derived cans on any significant scale.

3 BACKGROUND RESEARCH

Despite only partial technical success by the detinning industry in achieving an economic solution to the large scale detinning of refuse derived cans, AMG continued to support and initiate research programmes in this area.

One of the major reasons for the emphasis on the development work has been the steady decline of the volume of industrial tinplate available for detinning. The decline is due entirely to improvements in the efficiency of both the manufacture of tinplate and of tinplate cans. The utilisation of a refuse derived feedstock to supplement the prime industrial grade scrap is an obvious objective given the requirement to fill existing detinning capacity.

Several internal and external research programmes were commissioned during the last decade. Perhaps the most notable was the creation of Material Recovery Limited in 1975 - a jointly owned company formed by Metal Box, British Steel and AMG.

Briefly, the objective of M.R.L. was to develop and assess the most technically efficient method of extracting cans from the refuse stream and to provide a product that could be effectively detinned. After meeting partial success, this programme was eventually assumed by M.R.C.P. Ltd. whose contributors/shareholders embrace all the major U.K. Can Manufacturers and their principal raw material suppliers.

It is important therefore to place the latest phase of used can development work in its historical perspective in order to assimilate and recognise the volume and breadth of research that has preceded this current stage of development.

4 METHODOLOGY

There are two basic methods of detinning tinplate scrap:

4.1 Chemical Leaching

The most commonly known and adopted method involves the dissolution of the tin from the tinplate using caustic soda containing a suitable oxidant.

Chemical leaching using caustic soda based solvents is the preferred route, as the caustic has no deleterious effect on the base steel and tin dissolved as sodium stannate is easily recovered as metallic tin. Several mechanical methods are employed to conduct the leaching operation - normally perforated baskets/drums or the novel continuous process developed by AMG. In all cases, the process is conducted at temperatures in excess of 98°C and for varying residence times dependent on weights charged.

The temperature /reaction time and liquor concentrations are critical parameters, control of which enable optimum removal of the tin layer and, more importantly, iron/tin alloy layer.

Properly controlled, such chemical processes enable residual levels of tin to be reduced to less than 0.02%.

The tin bearing solution from detinning is treated to remove impurities and then electrolysed by conventional electrowinning technology where the tin is plated out of solution, which after melting, produces ingot tin of 99.95% min. purity.

The tin depleted solution from the electrowinning process is adjusted by caustic soda, oxidant additions and is then re-used in the leaching process.

4.2 Electrolytic

The most commonly adopted electrolyte is based on caustic soda.

The process involves using the tinplate scrap as anode in an electrolytic cell.

The tinplate is generally packed into steel baskets connected to the anode busbar in a cell.

Electrolytes are typically 6 - 8% caustic soda and temperatures maintained at 70 - 90°C.

Tin is plated out on to steel cathodes, which are periodically scraped.

Both technologies require the scrap to be clean, free from contaminants and be of an homogeneous nature. If the scrap is to specification - and the vast majority of industrial (factory) tinplate scrap does meet that specification - tin recovery is maximised and the detinned steel is an ideal feedstock for steelmaking.

However, there are several inherent physical characteristics of tinplate cans extracted from the refuse stream that severely inhibit effective chemical detinning and have repercussions on both quality and process costs.

1. Dirt Contamination:
   Residual contents, paper, plastics, ash etc.
2. Aluminium Contamination:
   Principally bi-metallic cans.
3. Tramp Materials:
   Other non-can ferrous metals.

4. Surface Area:
   Limited flat surfaces, low density seams etc.
5. Can Type:
   Non-homogeneous, e.g. welded, three piece, two piece, bi metallic, aerosols etc.

To achieve the specification required by the detinner, the used cans have necessarily to be pre-treated prior to detinning. The cans need to be cleaned, the aluminium, paper and plastic removed, the tramp materials screened and the product homogenized (Preparation).

5 THE RECLAMATION PROCESS

The reclamation process embraces two quite separate technologies. Briefly, cans extracted from the refuse stream must first be prepared by specialised shredding and cleaning techniques to provide a suitable feedstock for subsequent detinning. Secondly, the detinning process itself must be amended to handle the unique nature of this material.

5.1 Preparation

Much of the research carried out by AMG in the last ten years has been centred on developing a single treatment and cost effective process which eliminated deleterious elements and at the same time provided a suitably consistent product for detinning.

The heart of the AMG process is its world-patented Cutler shredder. The shredder was developed by AMG in the U.K. following exhaustive trials of alternative technolgies which, after rigorous assessment, did not meet the demanding technical and commercial criteria needed to provide an economic solution.

The Shredder, coupled with cyclonic air extraction, magnetic separation and trommel screening provides that suitably consistent product.

Essentially, the shredder employs the principle of a contra-rotating cage and impeller. As the used cans are fed into the shredder, the cyclone created impels the cans towards each other. On colliding, any paint and lacquer are stripped and scratched and the fragments shear each other thereby liberating the various fractions, and the two major contaminants of dirt and aluminium may be readily removed by subsequent air classification and magnetic separation.

In summary, the shredder and ancilliary equipment provides the following advantages in the preparation of used cans:-

1. A single continuous treatment process.
2. Removes 98% of dirt contamination.
3. Removes 99% of aluminium.
4. Separates tramp materials.
5. Provides homogeneous detinnable product.
6. Low running cost and simply maintained.
7. Relatively low capital cost.

5.2 Detinning

Alongside the development work specific to the preparation of used cans prior to detinning, further research was initiated into detinning methods. The detinning processes employed by AMG were not entirely suited to the detinning of the prepared used cans. The non-metallic impurities introduced by the used can feedstock rendered the use of the chemical leaching route unsuitable both on a technical and cost basis.

The electrolytic route formerly employed by AMG in the early 50's was adopted for use on the new type material.

The shredding action produces a dense tinplate pellet which is physically very different from the light industrial tinplate scrap produced from the manufacture of cans whether two piece, three piece or aerosols.

An earlier detinning technique, once employed by AMG but discarded in the 1960's, was totally re-worked and adapted to detin the new type of material.

Not only were the chemical and electrical efficiencies improved to provide a faster detinning cycle and more effective tin winning, but also mechanical handling techniques were transformed to register significant productivity gains.

An added advantage of this detinning system is that it is also capable of detinning conventional industrial tinplate scrap.

6 THE DETINNING OF USED CANS - THE EXPERIENCE

Toward the end of 1984 a 30,000 tonnes per annum dual purpose plant employing the new technology was built at AMG's site at Hartlepool U.K. The first phase of the used can preparation plant was built in 1985 and in 1986 the second phase was commissioned which was officially opened on 30th September by John Butcher, M.P., Parliamentary Under Secretary of State for Industry. The preparation plant has a capacity of 15,000 tonnes of used cans per annum, equivalent to 375 million cans per annum and the plant is currently operating at the rate of 1.25 million cans per day.

The preparation and detinning plants have met their design specification whilst tin yields, operating costs and detinned steel quality have achieved their targets. A second plant with a capacity of 20,000 tpa was commissioned in 1992 at the company's plant in Llanelli.

The technology developed in the U.K. has been successfully employed in the U.S. The first Plant at Pittsburgh and a second in Minnesota are now operational. Greece and Yugoslavia are the first European countries to adopt the technology and AMG have plans for plants in other European countries. A 40,000 tpa plant utilizing the technology was commissioned during 1992 in Australia in conjunction with Edward Levy and BHP.

The overall energy efficiency of the process has been audited by the Directorate for Energy of the E.E.C. which has reported significant energy saving against any other options.

The only constraint on the proliferation of the process is the availability of raw materials. Almost uniquely among recyclables there is no market constraint on the finished products.

There is an increasing awareness of the need for conservation and this is resulting in pressure for various types of legislation designed to conserve natural resources and reduce the volume of waste for ultimate disposal. I would urge legislators to think and consult very carefully. There are many examples of well-meaning but ill-considered legislation designed to protect the environment which in practice have proved to be counter-productive by distorting established patterns of movement of recyclable materials.

The plant and technology that I have described was designed and initially operated utilizing magnetically separated raw materials.

It must be recognised that there are many areas both in the UK and abroad where, for a number of reasons, magnetic extraction is not practical. In these areas a rapid and self accelerating growth in consumer aided collection schemes is taking place.

The plant is equally efficient in processing metals recovered by these schemes which are providing an increasingly important source of raw material and a response to the public's obvious concerns with recycling and the environment.

Why Recycle?, Rainbow (ed.) © 1994 Balkema, Rotterdam, ISBN 90 5410 367 1

Beneficial re-use of used paper products

Leslie Webb
Envirocell, Ashtead, UK

ABSTRACT: The environmental effects associated with three scenarios (recycling, incineration and landfill) for re-using used paper products are considered qualitatively in terms of selecting raw materials with beneficial characteristics. It is concluded that the major paper components have a beneficial potential in each scenario with the minor components being responsible for most of the potential problems.

1 INTRODUCTION

Used paper products may be re-used beneficially in a number of ways, but the chemical composition of each product dictates the value in each route and the associated resource inputs to gain that value. As the papermaker has no control over the actual routing of the used product, it is important that all papers are manufactured using raw materials and processes that maximise all aspects of re-usability consistent with product functionality.

After an introductory review of papermaking raw materials, this paper will evaluate the composition of several major paper grades in terms of their re-use potential in either recycling into more paper products or incineration judged against landfill as the normal disposal standard.

2 PAPERMAKING RAW MATERIALS

2.1 Fibres

Finished papers represent a diverse group of products based on cellulosic fibres, the latter accounting for close to 100% of the product's weight in some grades such as newsprint and tissue, but only about 50% of the product's weight in some types of coated paper. The natural fibres provide the paper's basic strength, whilst other raw materials introduce other key characteristics such as improved opacity, resistance to liquids, strength under wet conditions, colour, brightness, etc.

The raw fibre is largely sourced from managed forests, the main wood pulp-producing regions being the Nordic countries, Canada and the USA. Non-wood cellulosic fibres such as cotton, straw, flax, etc are mainly used in developing countries. Used papers (wastepaper) are an important secondary resource in many countries, particularly those such as the UK with limited virgin fibre sources.

The pulps used in papermaking are produced by pulping and bleaching the raw fibres in one of two broad categories of process:

* mechanical processes in which most of the wood substances are retained in the papermaking pulp and thus have a high fibre yield of 85-95%. These pulps may be bleached with chemicals such as hydrogen peroxide or sodium hydrosulphite and are mainly used in short-duration products such as newsprint and magazine papers.
* chemical processes in which much of the original lignin and some of the hemi-celluloses are

dissolved in the pulping liquor and thus have a low-moderate yield of 40-80%. Unbleached chemical pulps are mainly used in packaging papers/boards. These pulps may be bleached with a number of possible chemicals (eg chlorine, chlorine dioxide, oxygen, ozone, etc), the bleached pulps being used in products such as the typical range of office papers, some tissue grades, etc.

Used papers are recycled largely for their fibre content, but other components could also be re-utilised depending on whether the used paper is being recycled into a similar paper grade or a different one. There is a similar division of wastepaper re-pulping processes into mechanical and chemical variants. Mechanical processes are mainly used for recycling used packagings into similar products, whereas chemical (deinking) processes are used to clean printed papers for re-use in grades such as newsprint, other printing papers and tissue/towel.

In all of these processes, there are significant inputs of other resources (chemicals, energy, water) in order produce the paper-making pulp.

2.2 Mineral fillers/pigments

After fibres, minerals are the largest tonnage raw materials used across the whole paper industry, but some products (newsprint, tissue/towel, most packagings) contain little or none at all. Mineral fillers are used as cheap extenders in printing papers to improve opacity and surface smoothness and more expensive minerals are used as pigments in the coating of some papers.

The main two minerals used as both fillers and coating pigments are kaolin clay and calcium carbonate. Titanium dioxide is also used to improve opacity/ brightness, notably in light-weight papers. Filler levels of 20-30% are not uncommon in many uncoated grades and coating can increase this to around 50%.

2.3 Other chemicals

The next most significant raw material is starches, which are used to improve the dry strength of some papers, eg printing papers, tissues and wastepaper-based packagings. Starch levels up to about 10% are possible, but are more typically 3-5%. Other chemicals (eg some natural gums, polyacrylamides, etc) may also be used for this purpose, but starches are easily the most prevalent.

Papers used under wet conditions (eg some packagings, towels, etc) have to be reinforced chemically so that they do not simply fall apart when being used. This is achieved by incorporating chemicals such as urea-formaldehyde resins or polyamine-epichlorhydrin resins at levels up to about 1% of the paper's weight.

In order to resist water penetration, some papers (eg wood-free printings/writings) have to be sized with hydrophobic chemicals such as natural rosin (usually used with an aluminium compound) or synthetic materials such as alkyl ketene dimer. These chemicals are typically present at low levels (up to about 0.5%) in the finished paper.

Whilst the above materials are used to introduce some essential functional effect into the paper product, dyes are used for purely decorative purposes. A wide range of chemical types are used and, whilst the addition levels are normally low, they can approach starch levels in highly-coloured papers. Fluorescent brighteners are chemically similar to dyes, but are more functional as they are used in white papers to overcome the yellow/brown hue from low levels of residual lignin in the pulp.

Other materials that may purposely be present in some papers include preservatives, softening agents and coating binders. Others may be present incidentally through their use to control some aspect of the papermaking process, eg biocides for slime control, polymeric flocculants for retention/ drainage improvement, defoamers, etc.

3 RECYCLABILITY OF PAPER COMPONENTS

3.1 Recyclability criteria

Theoretically, all the materials added to the paper by the paper mill are recyclable provided that they can be effectively retained in a functional form in the paper the next time around. The poor recyclability of some papers is usually caused by materials added to the paper in later operations such as converting (added adhesives) and printing (inks). Nevertheless, all the materials incorporated in paper products should be evaluated for their recylability characteristics in terms of:

* for papermaking materials, ease of retention in a functional form without causing process problems and
* for non-papermaking materials, non-interference with papermaking process and/or ease of removal from recycled pulp in terms of minimising inputs of chemicals and energy. (This aspect will not be covered in this paper.)

3.2 Fibres

All the fibrous materials in used papers can be retained and have value in a recycled product, but this varies between different fibre types. The value of mechanical pulp fibres are maintained over several papermaking cycles, whereas chemical pulp fibres lose their initially-high strength quite rapidly (Howard, 1991). In practice, the actual effect of recycling on fibre properties depends on the detailed fibre history involving factors such as the original papermaking chemistry (eg pH), paper drying and paper age (Cathie, 1993). So far as it is practicable and economic, papermaking should be carried out in a way that maximises fibre recyclability, but the somewhat intractable problem is that the most damaging step is the actual drying of the paper web.

It is thus likely that all fibres naturally lose some of their papermaking value on continued recycling. This deterioration can be ameliorated in any of three ways:

* by reactivating the fibres, eg by mechanical treatment, but this has undesirable effects on other papermaking properties such as drainage, or by chemical treatment under alkaline conditions
* by introducing other materials into recycled grades, eg starch as already practised for corrugating medium to achieve strength levels comparable to that of virgin fibre-based products
* by using at least some virgin fibres.

It is now widely accepted that an input of virgin fibre to the paper cycle is essential at the very least to replace fibres that leave the paper cycle deliberately through inherent non-recoverability of some paper products or incidentally through inadequate recovery infrastructure or inadequate market outlets for recycled products. However, it is not known with any certainty how the introduction of virgin fibres should be engineered in order to maximise their recyclability and sustain a high level of recycling, whilst minimising other resource inputs.

Clearly, virgin fibres should be utilised in products that are potentially recoverable (eg newspapers, magazines), even if these are not actually yet recovered with a high efficiency (eg office papers), but should be discouraged in products that are inherently not recoverable for recycling (eg towel/tissue). Non-recoverable products should be made with recycled fibre provided that this meets the normal product specification. Additionally, the product composition should facilitate, or at least not negate, beneficial utilisation in its normal disposal route (ie during sewage treatment for toilet tissue).

The papermaking potential of recycled fibres is maximised if they are either recycled into the same or similar paper grade or downcycled into a product with less onerous fibre requirements. Fibres can be upcycled into paper grades with more searching fibre requirements, but this would require significant additional resource inputs (chemicals, energy) which may exceed the inputs to produce such a pulp from virgin fibre sources. Another factor for consideration is that upcycling (and downcycling) is less likely to be able to utilise the other components in the used paper.

3.3 Mineral fillers

Unlike papermaking fibres which are subject to chemical changes due to biodegradation and other processes, mineral fillers are more stable to the general range of conditions encountered during recycling. Arguably, it is more important for minerals, despite their cheapness in straight economic terms, to be recycled in a beneficial way as they are a non-renewable resource, albeit one with considerable reserves in many cases.

Minerals are eminently recyclable provided that filled papers are recycled into paper grades that are able to utilise the functional properties of the fillers. Genuine recycling into the same or similar paper grade will thus maximise the overall recylability of a filled paper and reduce the raw material requirements of the recycled grade. At the present time, the filler content of recycled papers is not generally utilised as filled papers such as magazines tend to be downcycled into filler-free grades such as newsprint and tissue/towel. The filler then accounts for a major part of the residues from the deinking process, but the deinking sludge can be utilised beneficially (eg by land application) (Stringer, 1993). The re-use of fillers from recycled papers is likely to improve as more filled grades (particularly office papers and magazine papers) are made with at least some recycled material.

There are differences in the recyclability of the two main papermaking fillers, clay and calcium carbonate. Clay is inert chemically under all deinking and papermaking conditions, but calcium carbonate can be dissolved under acid conditions and is thus effectively destroyed as a subsequent papermaking material. Acid conditions do occur in deinking processes and some paper products are still made under acid conditions although there has been a major shift away from acid to neutral/alkaline papermaking over the last 20 years. This trend is driven by economic and quality factors, but it can be argued (Webb, 1993) that other environmental advantages of calcium carbonate over clay outweigh this slight constraint in terms of paper recyclability.

Even when the fillers are not recovered for recycling into the recycled paper, however, there are possible benefits from their presence. In the flotation deinking process, clays are known to improve the efficiency of ink removal and the controlled dissolution of calcium carbonate provides a "free" source of calcium ions that are needed to react with added soap for ink collection and removal.

3.4 Other chemicals

At the present time, most other materials in used papers for recycling are not utilised in any positive way. They may be retained incidentally in the recycled product, but it would be difficult, if not impossible in most cases, to ascribe any particular functional effect to the retained material. Most of these materials are more important in terms of their potential for causing some degree of difficulty in making paper from used products containing them, viz:

* soluble starches increase slime growth and related problems and contribute to wastewater BOD/COD
* sizes may hydrolyse thus causing deposits and interfere with fibre bonding
* wet strength agents make the paper difficult to repulp
* dyes cause shade variations and would thus have to be removed by bleaching for making white papers
* coating binders can cause pitch deposits.

As starches may be present at significant levels in many used papers, their recyclability is more important than the other materials listed above. The recyclability of starches depends on their chemical character, notably if and how they have been modified by the starch supplier. Unmodified starches are still the most prevalent type used in papermaking and are largely added to the surface of the paper after the fibrous web has been formed. When such papers are recycled, the starches are released into water and not retained with the paper. By contrast, starches containing cationic groups are adsorbed by the recycled fibres and retained with the paper (Mayatt, 1993). As a consequence of their good adsorption onto fibres, these starches are already widely used for addition to the fibres before the paper sheet is formed, but the addition rate for this application is low (<1%). Their use at the paper surface is limited by their higher cost compared to unmodified starches, but this can be compensated for by the overall benefits related to their better retention on recycling (Webb, 1992).

4 INCINERATION OF USED PAPERS

Although little used at present in the UK for dealing with domestic and commercial waste, the use of incineration is increasing rapidly due to pressures on landfill and the subsidies available for power thus generated from a non-fossil fuel source. Within mixed waste, used paper products represent an important source of combustible material that is itself obtained from a renewable resource and therefore more sustainable than the other major combustible material in solid waste - plastics of various types. Several recent studies (eg Hamm, 1993) have shown that increased incineration of wastepaper could make a substantial contribution to reducing CO_2 emissions from fossil fuel sources.

The energy value of the principle organic components of used paper (mainly cellulose and starch) is about 20 GJ/tonne or about half that of coal/oil due to their more oxidised form. The organic content of papers varies from about 50% for some coated papers up to 100% for tissue/towel. Although the presence of mineral fillers detracts considerably from the combustibility of some used papers, the calcium carbonate has some value in neutralising the emission of acid gases such as SO_2, albeit then releasing a corresponding quantity of CO_2. The clay filler has no comparable benefit and, like the calcium carbonate/sulphate), contributes to residual ash for disposal (or re-use).

There are some sources of sulphur and nitrogen compounds in most papers (eg sulphur in papermakers's alum and some dyes, nitrogen in polyacrylamides and wet strength agents), but the levels are generally low. Chlorine-containing chemicals are not widely used although chloride is usually the counter ion for most cationic polymers used. The chlorine content of virgin pulps that have been bleached with chlorine compounds using current technologies is also low (<0.1%). Paper products within mixed waste are thus not a major source of pre-cursors of emissions of sulphur-, nitrogen- and chlorine-containing substances.

The minor constituents of wastepapers are mainly organic and will thus contribute proportionately to the net energy yield from incineration. It has been estimated that incineration of all UK solid waste could generate about 10% of UK primary energy requirements of which non-recycled wastepaper could contribute 10% (ie 1% of total UK requirements).

5 LANDFILL OF USED PAPERS

As little solid waste is incinerated in the UK, the majority of used papers (about 6M tonne p.a.) is still landfilled. The driving forces for utilising wastepaper in other ways is usually justified on the grounds of reducing impacts associated with landfill. Evaluation of the landfill impacts of paper components is thus important in relation to judging the soundness of other routes and in absolute terms in view of its current prevalence as a disposal route. It should be noted that the landfill impacts do depend markedly on how the landfill is operated in terms of handling its products/emissions. Some of the characteristics of paper components, notably biodegradability and toxicity, are relevant to other waste utilisation options, such as composting and anaerobic digestion, as well as to landfill.

The fibrous components of used papers are inherently biodegradable, but the rate of biodegradation depends on environmental parameters (pH, moisture, etc) and on the lignin content of the fibre. The high lignin content of newsprint is probably one of the factors in the oft-reported longevity of intact newspapers in some landfills, although this will also be influenced by environmental parameters such as moisture. Starches behave in a similar way, but are broken down more quickly and thus contribute more to the organic strength of early leachates. This biodegradability of the major part of paper products is a problem or a benefit depending on whether or not the landfill gas is collected and, if it is, whether it is simply flared-off or burnt for energy recovery.

The behaviour of the mineral fraction of wastepaper in landfills depends on mineral type. Clays are effectively inert, as is calcium carbonate under neutral or alkaline conditions. However, landfills tend to become acidic due to biological activity and to the possible input of acidic wastes. Calcium carbonate thus has a useful role in neutralising such acidity which would be beneficial in terms of maintaining neutral conditions for optimum biological activity and for arresting dissolution and movement of heavy metal ions. Some of the latter, albeit at low levels, can originate from some wastepaper components such as inks and dyes.

Many of the other minor components of used papers are not biodegradable and will thus persist indefinitely in landfills. Rosin size is slowly biodegradable under aerobic conditions, but some of the component acids (originating from wood) are toxic at low concentrations. Hydrolysis and breakdown of rosin releases aluminium ions, which are then likely to be either precipitated or adsorbed on other particulate solids. Alkyl ketene dimer size is non-biodegradable.

Once cross-linked, wet strength agents in paper are insoluble and unlikely to be attacked chemically or biochemically and will impede to some degree the degradation of the surrounded fibres. Water-soluble polymers such as polyacrylamides are also non-biodegradable, but their charged character limits their potential for translocation via leachates.

The biodegradation, toxicity and bio-inhibition characteristics of dyes have been widely studied (Brown and Anliker, 1988). Most dyes (and the related fluorescent brighteners) are not biodegradable under aerobic conditions, but some (eg the common azo dyes) may be degraded anaerobically to aromatic amines. Although some dyes are quite toxic and/or inhibitory (to aquatic species), the potential for toxic/inhibitory effects in landfills is limited by their generally-low concentrations and their substantivity to particulate matter.

6 CONCLUDING REMARKS

It is evident that a high proportion of the materials present in used papers can not only be recycled effectively into further paper products, but can be beneficially utilised in other ways. The papermaker can maximise the recyclability and the benefits from wastepaper incineration or landfill through a process of environmental screening of all raw materials.

The dominant fibrous component (of whatever type) is valuable in most scenarios, but there are important differences between the two main types of mineral filler and different types of starch. Although the minor components of used papers exert more subtle effects in recycling and re-use, their largely adverse effects can still be minimised by careful selection.

This paper has only looked at the pros/cons of papermaking raw materials in one part of the paper cycle and one that has not, until recently, received much interest or attention from the papermaker. Identifying the "best" raw materials and processing conditions to minimise environmental impacts over the whole paper cycle requires use of the developing tool of life cycle assessment (LCA). In the absence of rigourous LCA methodologies, the papermaker can still make valid decisions about optimal raw materials combinations using the qualitative approach used in this paper.

REFERENCES

Brown, D and Anliker, R 1988. Dyestuffs and the environment - a risk assessment" in "Risk assessment of chemicals in the environment", Royal Society of Chemistry, Ed. M.L.Richardson, Chapter 22, 389-413.

Cathie, KC 1993. Extending fibre life. Paper presented at Pira International/Paper Focus Seminar "Recycling paper means business", London, May 1993.

Howard, RC 1991. The effects of recycling on paper quality. Paper Technology, 32, 4, 20-25.

Mayatt, J 1993. "Cationic starches at the size press and cost effectiveness" Paper presented at Pira Conference "The chemistry of papermaking", Wilmslow, January 1993.

Stringer, D 1993. Agricultural land improvement. Paper presented at PITA Seminar "Towards zero waste: problems and opportunities", Bolton, May 1993.

Webb, LJ 1992. Secondary fibre and wet end chemistry. Paper presented at Pira Conference "The chemistry of papermaking", Wilmslow, January 1992.

Webb, LJ 1993. Guide to best environmental practice techniques for paper mills. Envirocell Report R4, 122 pages.

Why Recycle?, Rainbow (ed.) © 1994 Balkema, Rotterdam, ISBN 90 5410 367 1

Bleak Recycling Prospects for the UK

C.Wilkins
Stirling Recycling Ltd, UK

Waste paper is a problem. In todays market terms the UK cash value of the material does not justify the collection costs and certainly cannot rival the negative costs on the continent. There has to be a subsidy but who should pick up the bill ? A pragmatic approach starts with an international waste paper market dominated by the cost of landfill. In Scotland landfill is only a few pounds per tonne, while in Northern Europe it is around £100 per tonne.

In Germany, where there is little landfill capacity left, there is a very different policy to the UK: a levy is placed on all packaged goods and the revenues are used to fund the collection of packaging waste, thus giving German paper mills their raw materials free of charge and also producing an enormous surplus. The additional waste paper exported by Germany last year is estimated at 600,000 tonnes and this is projected to move up to 2.5 - 3 million tonnes within the next three years. When the same arrangement is made for newspapers and magazines the tonnage will increase by a further 2 million tonnes, making a possible 5 million tonnes/year available. This figure has to be put into perspective, it is more than one and a half times as much as the total collected in the UK in 1992.

It is a political decision as to whether a country recycles or not, because it comes down to public money or political will. However the benefits of recycling are very wide and apart, from the environmental ones, they include producing competitively priced raw materials which - in turn - helps manufacturing industry, employment and the balance of payments. If our competing nations put money into generating raw materials from waste then either we should as well, or be clear about the effects of opting out.

To take newsprint as an easily identifiable part of the problem, it quickly becomes waste that can then - given the appropriate treatment - be converted back into newsprint. In Germany 730,000 tonnes of additional recycling newsprint capacity are under construction but there are none in the UK, despite the fact that the UK imports well over 1 million tonnes/year. Germany will now become a net exporter of newsprint.

UK newsprint is estimated to use only around 20% British recycled fibre and some 2 million tonnes of usable material goes to waste ! This means that if the UK had 2 new machines (500,000 tonnes capacity) these, together with existing capacity, could satisfy 44% of consumption with recycled and that would use only 40% of the available post consumer waste paper. It would also nearly half our imports of newsprint.

The UK Government is demonstrating its support for building the newsprint industry. The Swedish company, SCA has been offered £20 million to build in Kent and North British Newsprint has applied for a similar type of assistance for its proposed mill in Scotland. This makes aid in the UK - more or less - internationally competitive but it does not help if the raw material costs are all wrong. In the UK paper mills have to pay the real cost, in Germany they get the raw material free.

The UK is a member of the EC. There is supposed to be a competition policy and either we should make the EC change or our system should change. The present UK policy is laissez faire: Government has set a target to recycle 25% of household waste by the year 2000. A significant part of this is the 2 million tonnes of old newspapers and magazines that are thrown away each year. To help achieve the objective the newspaper publishers (the active consumers) have agreed a voluntary target of using 40% recycled by 2000. However, due to the recession there is more supply than demand and newsprint prices are 30% below their 1989 level. The upshot is that, in commercial terms, it is not the time to bring on new capacity in the UK. Even though it is right in environmental and structural terms.

It would be nice to think that Central Government and the newspaper proprietors are being pro-active in the field of recycling and not just reactive. Unfortunately this does not yet seem to be the case but let us hope that mounting pressure from the environmental lobby and the fear of being swamped with German paper will encourage them to take a more dynamic approach. If they do not then neither the Government's nor the publisher's targets will be met, the waste paper industry will be wiped out and British newspapers will be printed on paper which is an export of some other country's landfill problem.

Why Recycle?, Rainbow (ed.) © 1994 Balkema, Rotterdam, ISBN 90 5410 367 1

Water recycling in manufacturing industries and water pollution control

C.R.Hayes
Water Quality Management Ltd, Buckden, Huntingdon, UK

N.J.D.Graham
Imperial College of Science, Technology and Medicine, University of London, UK

ABSTRACT: Many manufacturing companies are being pushed by the regulatory agencies (NRA, HMIP and water service companies) to achieve more stringent water quality standards. Such standards derive from EC directives and from the development by the NRA of Statutory Water Quality Objectives.

Tighter control of water pollution may be necessary: for surface water drainage, for industrial discharges to controlled waters, and for discharges to the public sewer. This will need an integrated water pollution control plan in many circumstances, and improved pretreatment of discharges may be necessary.

As the cost of discharging waste waters is noticeably increasing, it may be advantageous to link the more extensive pretreatment of discharges to possible opportunities for water recycling via supplementary treatment. The options for treatment are reviewed.

1 INTRODUCTION

For many years in the United Kingdom, water supply and wastewater disposal costs have been low, consistent with successive Government policies. Cheap water was feasible when the breadth and stringency of environmental regulation was limited.

Since the late 1970's, an increasing range of European Commission directives, international agreements and UK Government initiatives, have combined to impose numerous strict quality standards on water supplies and wastewaters. These are generally being implemented as short to medium term requirements, rather than long term goals, and are enforceable by UK law.

In consequence, the cost of water supply and wastewater disposal seems likely to double during the 1990's, ignoring inflation.

Regardless of these rising costs, many manufacturing industries now face far more demanding legislation, particularly in respect of pollution control, and must face up to greater risks of prosecution by the regulatory authorities, and to possibly severe fines or even imprisonment.

It is now time for many companies to confront this challenge, to protect both their business viability and their environmental reputation, these aspects increasingly being linked.

It is also the time for identifying the opportunities that may exist for minimising the cost impact of this recent environmental legislation.

Water recycling is one such opportunity.

This paper reviews the most relevant environmental legislation, provides illustrative cost analyses, outlines the range in water and wastewater treatment technologies that are available, and indicates how manufacturing companies will need to proceed if they are to identify their potential for cost saving from water recycling.

2 ENVIRONMENTAL REGULATION

2.1 *Public water supplies*

The water companies are required to provide "wholesome" water at the "time of supply". In general terms, wholesome water must comply with the quality standards set by the Water Supply (Water Quality) Regulations 1989 and associated Amendment Regulations. In certain cases, the water company may have secured a relaxation of a quality standard for a particular water supply zone; such relaxations are mostly subject to conditions and time limits.

The "time of supply" is taken to be the point of entry to the premises of the consumer. With a few possible exceptions, the consumer is responsible for water quality after the time of supply (eg such as in factory premises).

The water companies are subject to the scrutiny of the Drinking Water Inspectorate, in connection with water quality, with enforcement action possible by the Secretary of State. Local district and borough councils have a duty to keep themselves informed about the wholesomeness of water supplies in their area and must notify the Secretary of State of any concerns not dealt with satisfactorily by the water company.

It can be concluded that the quality of public water supplies is being controlled tightly to very high standards, which may exceed the needs of many of the industrial applications for which the water is used.

The maintenance of high water quality standards within industrial premises is required by the Food Safety Act 1990 for food preparation businesses, and by the public health legislation in connection with water used for drinking, cooking and personal washing.

Public water supply charges are increasing steadily, reflecting a major current programme by water companies to improve treatment and distribution facilities, necessitated by the more stringent implementation of water quality standards. These charges are regulated by the Director General of Water Services.

As a general guide, public water supplies cost about 60p per cubic metre.

2.2 *Private water supplies*

The quality of private water supplies must comply with the Private Water Supplies Regulations 1991. There are many similarities to the regulations governing the quality of public water supplies.

Except for very small supplies, a licence to abstract water is required from the National Rivers Authority (NRA). Abstraction charges are levied by the NRA.

In most cases, the local district and borough councils administer the water quality regulations, having powers to relax water quality standards in certain cases and powers to seek improvement. When an Improvement Notice is served by a local council, the council must specify the steps to be taken and their timescale. The owner of the private supply, upon which a Notice has been served, has the right of appeal to the Secretary of State.

As a general guide, the NRA's abstraction charge for a private supply is about 1 or 2p per cubic metre. This cost does not include for any pumping, treatment or distribution costs that may be incurred, nor for any monitoring charges from the local council.

2.3 *Wastewater discharges to environmental waters*

In England and Wales, discharges to environmental waters are controlled by the NRA, having taken over such functions from the former Regional Water Authorities. In Scotland, control is exercised by the River Purification Boards, and in Northern Ireland by the Department of the Environment (NI).

In cases in England and Wales where "red-list" dangerous substances are involved, pollution control is regulated by Her Majesty's Inspectorate of Pollution (HMIP).

A system of Statutory Water Quality Objectives (SWQO's) is being developed by the NRA which will define the quality needs for all controlled waters, including rivers, lakes, estuaries and coastal waters. The NRA will then be able to identify more clearly the conditions that need to be imposed upon all discharges to a particular controlled water. It seems that SWQO's will not be implemented until the year 2000 in many cases.

In the mean time, reliance is still being placed upon quality objective schemes for environmental waters, that are less precise. Nonetheless, it is clear that any polluting discharge to an environmental water is an offence, unless that discharge complies with a formalised consent.

Arising from the Water Resources Act 1991, a pollution offence may be heard in a Magistrate's Court, which is empowered to levy a fine not exceeding £20,000 and/or a term of imprisonment not exceeding three months (plus costs, plus further fines for a continuing offence). A serious pollution offence could, however, be taken to the Crown Court which can levy an *unlimited* fine and/or impose a term of imprisonment not exceeding two years.

In addition to the body corporate, Company directors and any other person carrying relevant specific responsibilities are liable personally to prosecution for a pollution offence.

The NRA has powers to serve a Prohibition Notice on any body or person if necessary to exert pollution control.

The stronger enforcement of pollution control by the NRA may force a manufacturing company to exert much tighter control over its site (eg reducing the risk of pollution to surface drainage) and to significantly improve the pretreatment of any regular discharges.

The NRA charges for consented discharges, a typical unit cost being between 1 and 10p per cubic metre.

2.4 *EC directives*

The three most relevant directives are considered to be:

1. Council Directive of 18 July 1978 on the quality of fresh waters needing protection or improvement in order to support fish life (78/659/EEC).
2. Council Directive of 4 May 1976 on pollution caused by certain dangerous substances discharged into the aquatic environment of the Community (76/464/EEC).
3. Council Directive of 21 May 1991 concerning urban waste water treatment (91/271/EEC).

The "fisheries" directive will form part of the basis for SWQO's. To date, its impact has been minimal as the great majority of water stretches that were "designated", under the directive, comply.

The "dangerous substances" directive is supported by "daughter" directives, each of which specify Environmental Quality Standards (EQS's) for receiving waters, and industrial emission standards. In the UK, these directives have been implemented via EQS's not emission standards. However, HMIP is seeking to limit industrial emissions by the implementation of "best available techniques not entailing excessive costs" (ie "BATNEEC"). The EQS's will be incorporated into future SWQO's by the NRA. In the mean time, the NRA is already using the directives' EQS's to set environmental objectives and, in turn, discharge consent conditions.

The "urban wastewater" directive requires the provision of minimum sewerage and sewage treatment, by specified dates, dependent upon the size of the conurbation served (ie population).

This directive is requiring, over the next five to twelve years, the provision of:

1. Primary treatment of all sewage which currently is discharged in crude form;
2. Secondary biological treatment of sewage, in many cases, which currently only receives primary treatment or is discharged in crude form;
3. Nutrient removal from sewage discharges to "sensitive" environmental waters.

The provision of more extensive sewage treatment will affect the cost of some industrial effluent discharges to the public sewer, as discussed later.

2.5 *International initiative on the North Sea*

The UK Government was a signaturee to the Final Declaration of the Third International Conference on the Protection of the North Sea, 1989.

This Declaration included the agreed intention for all the Member States involved to reduce by at least 50% the inputs of "Priority Hazardous Substances" to the North Sea from rivers and estuaries, and that all discharges should receive at least primary treatment. The list of "priority hazardous substances" includes all dangerous substances specified by the EC "dangerous substances" directives and all UK Government "red-list" dangerous substances. However, it also includes additional substances, the most significant to industrial discharges probably being copper, zinc, lead, arsenic, chromium and nickel.

The NRA has consequently been applying tighter metal limits on discharges from sewage treatment works, which in turn has led to tighter metal limits on trade effluent discharges to sewer. Discharges direct to an environmental water have also been subject to tighter controls.

2.6 *Trade Effluent Control*

In the context of this paper, "trade effluent" is defined as the discharge of wastewater to the public sewerage system.

Trade effluents are treated and disposed by the water service companies in England and Wales and by the Regional Councils in Scotland. These service providers also act as regulators and have enforcement powers to control both the quality and quantity of trade effluent.

The breach of a trade effluent consent can lead to prosecution, most normally at a Magistrate's Court.

Stricter consent conditions are being sought due to the pressures from the more onerous environmental standards that are being implemented. The level of enforcement has also tightened; ie water service companies are less willing to ignore the breach of consented conditions. In consequence, it may be necessary for a manufacturing company to substantially improve any pretreatment plant that it operates, or install such plant for the first time.

The installation by water companies of new primary and new secondary sewage treatment at many locations, to comply with the EC urban wastewater directive and the "North Sea Conference Declaration", will greatly increase many manufacturing company's trade effluent costs. An illustrative example is provided later.

The service providers are entitled to seek capital cost contributions in cases where a trader's activities lead to an increase in trade effluent volume or polluting load beyond that specified in a consent or agreement. This aspect needs careful attention.

Any dispute with the service provider over proposed changes to a trade effluent consent or agreement will be determined, via an appeal procedure, by the Director General of Water Services. Any appeal must have a technically sound basis.

2.7 *Implications of the new Environmental Regulations*

These can be summarised as:

1. The costs to a manufacturing company of failing to achieve pollution control objectives can be high via punitive fines.
2. Company Directors and their staff can be personally liable to prosecution and imprisonment.
3. The pollution control objectives sought by the various regulators are becoming more stringent and demanding.
4. These pressures will require tighter control over water at higher cost.
5. Water is now a much more important and expensive commodity.

3 OPTIMISATION OF WATER COSTS

3.1 *General Considerations*

Optimisation of water costs will *ONLY* be achieved if sufficient technical knowledge is available. Too often, the necessary detailed appreciation of water issues is absent, because as a low cost item in the past, water has been taken for granted.

It is worth noting that the environmental regulators are not infallible and that much of the new regulatory regime is still being developed. For these reasons, there may well be scope for some negotiation over regulatory controls to minimise their cost impact.

The premise of this paper is two-fold:

1. A manufacturing company is likely to have to incur both capital and revenue expenditure in achieving higher quality effluent, and in some cases higher quality water supply, particularly in the case of supplies owned by the company.
2. It is but a relatively small step for a Company in this position to consider both water economies that may be achievable within production, and to consider additional "tertiary" treatment to achieve an "effluent" quality suitable for recycling back for production use.

Regardless of pollution control needs, recycling of water may be cost effective anyway. In all cases, the actual savings, and their cost of achievement, will be site specific.

A simplified illustration is described below, with reference to the commonly used Mogden Formula for trade effluent charging.

3.2 *The Mogden Charging Formula*

The following formula is used typically by water service companies:

$$C=R+V+\frac{Ot}{Os}B+\frac{St}{Ss}S$$

Where:

C = Total charge in pence per cubic metre.

R = Reception and conveyance charge per m^3 of trade effluent.

V = Preliminary treatment charge per m^3 of trade effluent.

B = Biological oxidation cost per m^3 of settled sewage.

S = Treatment and disposal cost of primary sludges per m^3 of sewage.

Ot = The Chemical Oxygen Demand (COD) in mg/l of the trade effluent after settlement of one hour at pH 7.0 or at the pH of the mixed sewage.

Os = The COD in mg/l of average strength settled sewage.

St = Total suspended solids content of the trade effluent at pH 7.0 in mg/l (or at the pH of the mixed sewage).

Ss = Total suspended solids content of average strength crude sewage in mg/l.

Typical present day costs for the cost factors are:

R = 12p per m^3
V = 10p per m^3
B = 11p per m^3
S = 7p per m^3

Average strengths are typically:

COD = 420 mg/l
TSS = 370 mg/l

Thus for average strength trade effluent, the total treatment and disposal charge made will typically be 40p per cubic metre.

For a company using public water supplies and discharging its waste water to the public sewer, the typical total cost is therefore 100p per cubic metre.

For comparison, a company using its own water supplies and discharging its waste water directly to a river, will typically bear a total cost of less than 10p per cubic metre *PLUS* the company's own treatment and pumping costs where applicable. In practice, the stringency of the regulations governing private supplies and direct environmental discharges may be such that the company's own costs exceed 90p per cubic metre, in which case, public water supplies and sewer discharge become more financially attractive. However, these latter charges are increasing rapidly and so the economic balance for water use may change.

3.3 *The case for pretreatment and recycling*

As over 90% of wastewater discharges from manufacturing companies are to the public sewer, the case for pretreatment of trade effluent will be considered.

The assumed "current" case is:

1. Annual volume discharged = 300,000 m^3.
2. Average COD and TSS strengths apply.
3. Primary treatment only is provided by the water service company.
4. All water used is from the public water supply and there are no losses within the company's site.
5. The company does not provide any pretreatment for its wastewater.

The annual external cost would be £267,000.

The assumed "new" case is as above except that the water service company, following the provision of secondary biological treatment, now makes a COD charge.

The annual external cost would be £300,000.

If a pre-treatment plant was installed by the company to reduce both suspended solids and COD by 90%, assuming the "new" water company position of including a COD charge.

The annual external cost would be £251,000.

If additional tertiary treatment was installed by the company such that 50% of water use was recycled, and assuming the previous case of pretreatment by the Company.

The annual external cost would be £126,000.

Obviously, the company will incur both capital and revenue costs in the provision of pretreatment of wastewater and in the provision of tertiary treatment to facilitate recycling.

Individual company circumstances will determine the financial feasibility of the greater pretreatment of wastewaters and recycling, but it must be appreciated that the financial balance continues to shift in this direction as external charges increase.

4 POSSIBLE TREATMENT NEEDS

4.1 *Public Water Supplies*

Water treatment to facilitate process use commonly entails further purification by ion-exchange and/or reverse osmosis. For surface derived waters, prior organic scavenging may be required. Dosing of corrosion inhibitors is also often necessary in high temperature applications.

In large factory circumstances, microbiological quality control may require the boosting of chlorine residuals. Alternatively, disinfection by UV irradiation can be considered to minimise the growth of heterotrophic bacteria (eg as may cause food spoilage or in hospital environments).

It should not be necessary to instal treatment to achieve "wholesomeness" standards as these aspects are being dealt with by the water suppliers.

4.2 *Private Water Supplies*

In addition to the factory specific issues discussed for public water supplies, a company may need to install treatment to achieve "wholesomeness" standards, in relation to water used for drinking, cooking and washing (eg in canteens), and for water used in food production.

In these circumstances, the most common requirements for treatment will be:

1. Disinfection to over-come bacteriological contamination, by chlorination or UV irradiation.
2. Nitrate removal by ion-exchange or reverse osmosis.
3. Iron and manganese removal by oxidation and filtration.
4. Solvent removal by air-stripping or granular activated carbon.
5. Pesticides removal by granular activated carbon, possibly with ozonation.

4.3 *Pre-treatment of trade effluents*

Specific needs will clearly depend upon the activities of the individual manufacturing company. Such needs will be defined by the trade effluent agreement or consent with the water service company.

It should however be borne in mind that trade effluent consents require a company to identify any new components in their discharge which may lead to further specific consent conditions being imposed.

Common requirements for pre-treatment will be:

1. Solids reduction by settlement or flotation, with or without enhancement by chemical coagulation.

2. COD reduction by biological systems such as the activated sludge process and biological aerated filtration.

3. Oil and grease removal by interceptors or flotation.

4.4 *Recycling*

To determine any needs for tertiary treatment, the individual company must first specify the quality objectives for the water to be recycled. Such objectives will depend on the sum of individual process requirements, and consideration must be given to microbiological quality from a Health and Safety view-point.

Requirements for tertiary treatment could include:

1. Further removal of solids (ie polishing) by sand filtration.

2. Organics removal by granular activated carbon, possibly including ozonation.

3. pH adjustment by acid or alkali dosing.

4. Purification by ion-exchange or reverse osmosis systems.

5. Disinfection by chlorination or UV irradiation.

6. Polishing by ultrafiltration.

4.5 *Sludge handling*

A consequence of many treatment processes is the production of sludge. Generally, it is cost-effective to dewater sludge prior to disposal, options including filter pressing and centrifugation.

Consideration must be given to the correct disposal of sludge, the commonest options being land-fill and agricultural land.

5 TREATMENT TECHNOLOGIES

There are a large number of treatment technologies available for removing water-based contaminants, whether these are in nature particulate, colloidal or in true solution. The selection of unit processes and their configuration depends on the nature of the raw water, the degree of treatment required, the production of waste concentrates and the relative costs involved.

The following is a brief description of some of the most common treatment processes used for improving water quality.

5.1 *Particle removal*

Particle-liquid separation is normally achieved by means of horizontal-flow settling tanks in which sufficient time is allowed for particles to migrate under gravity from the body of the liquid to the bottom of the tank. The separation efficiency is dependent on the size and specific gravity of the particles and for organic-type solids it is typical to allow a detention time of 4 to 6 hours to achieve a solids removal of 60 to 70%.

The efficiency of the settling can be improved by the use of lamella plates. These are suspended within the settling tank in modules such that the flow of water passes through narrow channels and this provide a large area and short settling distances for particles to accumulate on the plates.

Settling can also be improved by the addition of chemical coagulants. Such chemicals are able to bind particulate and dissolved impurities within a metal hydroxide precipitate ('floc'). The floc particles are usually sufficiently dense and large to settle rapidly. Prior to settling a period of mechanical or hydraulic mixing is required ('flocculation') in order to achieve floc growth.

If the particles, coagulated or not, are unable to settle sufficiently quickly then their separation may be effected by flotation. The most common form of flotation is Dissolved Air Flotation in which a dense cloud of very small air bubbles are released from the bottom, inlet end of a tank. These bubbles attach to the particles in the flow and lift them to the surface of the flow. The efficiency of particle removal is generally greater for flotation than with settling systems, but the former requires additional equipment, such as compressors, saturators and recycle pumps.

For removing relatively low concentrations of suspended particles in the flow the most efficient method is filtration. This involved passing the flow through a bed of granular media, usually sand, so that particles are captured and retained within the pores of the bed. To remove the accumulated material the filter beds are automatically cleaned, every one or two days, by back-flushing with air and water.

The following is an indication of the relative scale of capital costs for particle removal processes:

conventional settling tanks £600 per m^3/h

coagulation-assisted settling	£350 per m^3/h
flotation (not including flocculation)	£360 per m^3/h
rapid gravity filters	£660 per m^3/h
proprietary pressure filters	£800 per m^3/h

5.2 *Biological methods*

The stabilisation of moderate-to-high strength organic wastes is normally achieved by aerobic biological processes. These operate by establishing an optimum population of heterotrophic and autotrophic microorganisms within a reactor tank such that the organic waste can be adequately purified within the tank's flow detention time of 2 to 6 hours. Current processes are based on either hydraulically suspending the microorganisms in the body of the liquid - the activated-sludge process - or where the organisms are attached on the surfaces of submerged granular media - the biological aerated filter.

The activated-sludge process is a well-established method with the capability of producing an effluent quality acceptable for direct discharge to receiving waters from moderate strength influent flows. Alternatively, in its 'high-rate' mode it can provide an efficient partial treatment of high strength waste flows prior to discharge to sewer.

The biological aerated filter process is a relatively new technique that is rapidly achieving widespread use in view of its higher loading rate, versatility and competitive costs.

An indicative value of the capital cost of such biological systems is £2,000 per m^3/h flow.

5.3 *Organic compound removal*

Wastewaters may contain a range of immiscible and miscible organic substances. Immiscible substances like oils and light spirits can be separated from the flow by virtue of their lower density than water in units called interceptors. Such units are baffled chambers such that the floating oils are retained between the baffles and these can be removed periodically by mobile suction systems.

Low concentrations of miscible organic substances can be removed effectively either by air stripping, if the substances are volatile (eg. solvents), or by activated carbon adsorption, if they are non-volatile (eg. pesticides). Air stripping is normally carried out in enclosed towers packed with media designed to provide a large surface area of contact between the water and air, and the contaminated air is continuously removed by a forced ventilation system. The capital cost of air stripping is in the range of £100 to £800 per m^3/h.

Activated carbon is a very effective general adsorbent for non-volatile, non-polar organic compounds, partly as a result of its very large specific surface area. Granular beds of activated carbon (GAC) are typically housed in open tanks or pressure vessels, in a similar fashion to sand filters. The effective lifetime of the carbon is usually designed to be in excess of one year before the carbon needs to be removed and regenerated. Since the hydraulic loading rates of GAC beds are very similar to sand filter beds, the capital cost of such units is also similar (ie £600 to £800 per m^3/h).

5.4 *Membrane filtration*

For producing very high quality waters from poorer quality raw waters, various types of membrane filtration can be applied. This involves forcing water through a semi-permeable membrane by means of an applied pressure, in order to concentrate dissolved salts and organic compounds. In reverse osmosis (RO) a membrane with a very small pore size is used and this can achieve a very high separation of dissolved solids, including heavy metals and large organic molecules. However, this requires operating pressures in the range of 15 to 50 bar.

Ultrafiltration and microfiltration are similar in principle to RO, but the membranes have much larger pore sizes and operate at lower pressures. Ultrafiltration membranes reject organic compounds of molecular weight about 1,000, and operate at pressures of normally less than 5 bar. Microfiltration membranes have pore sizes typically in the range of 0.01 μm to 12 μm, and reject colloidal and suspended material at operating pressures of 1 - 2 bar.

5.5 *Ion exchange*

Ion exchange resins are used either to remove specific ion contaminants, such as heavy metals or nitrate, or to demineralize the water. Ion exchange takes place as water passes through beds of synthetic exchange resins contained in pressure shells. When the capacity for exchange is saturated, the bed is taken out of operation and the resin regenerated with an appropriate chemical.

Plant costs are in the region of £500 to £700 per m^3/h.

5.6 *Oxidation and disinfection*

Strong oxidising chemicals such as chlorine, chlorine dioxide and ozone are useful for degrading biological and chemical (eg. ammonia) contaminants when used in the dose range of 1 to 20 mg/l. A high degree of microorganism inactivation can be achieved, particularly by ozone, in this dose range, as well as a substantial reduction in chemical and biochemical oxygen demand.

Whereas chlorine can be delivered to site and dosed directly, ozone must be generated on-site from air or oxygen and exposed to the water via contactor chambers. Thus, an indicative cost for chlorination equipment is £30 per m^3/h, whilst the equivalent value for ozone would be greater by a factor of 2 to 3.

An alternative, non-chemical, method of disinfection is the use of ultraviolet light (UV). In comparison with chemical means of disinfection, UV offers advantages of compactness and cost because contact tanks and chemical storage are not required.

6 APPRAISAL

6.1 *General Background*

Because water services have been cheap, many manufacturing companies have not considered it necessary to devote much (or any) of their technical resources towards water management.

We have commonly found that:

1. Trade effluent discharge volumes have been estimated from water supply meters, not measured, such that "lost" water still attracts a discharge cost.
2. The company relies totally on the water quality information provided by others, low sampling frequencies possibly leading to inaccuracies.
3. Knowledge of water use within production is often vague.
4. The company has relied totally on the pronouncement of regulatory bodies and not considered challenging or negotiating on regulatory interpretations.

The increasing cost of water services clearly warrants closer attention and appreciation by some companies.

6.2 *Expertise*

It can be appreciated from the preceding discussion that thorough appraisal, to identify optimum water management, will require extensive and broad expertise. Such expertise spans regulatory, scientific and engineering disciplines.

Many companies cannot cost-justify the specialist staff implied and will seek external consultancy support. Consultants must be selected carefully and preference given to those active in the full range of disciplines required.

6.3 *Appraisal Check-List*

In broad outline, the following wide ranging points may need to be addressed:

1. Can water use in production be reduced and how?
2. Can less expensive water supplies be obtained, particularly for non-potable use?
3. Is there scope for negotiation on regulatory controls?
4. Are the implications of present and future regulatory controls fully understood?
5. What is the risk of prosecution for failure to achieve regulatory standards?
6. Is the Company's "environmental image" of particular concern?
7. Are the circumstances of the Company such that better or new pretreatment of trade effluents should be considered?
8. What are the water quality objectives of any water recycled back to production?
9. Is sufficient information available on water flow, water quality and the structural condition of existing plant, drains, sewers, pipework, etc?
10. What new treatment plant will be needed and at what cost?
11. Would a site water management plan be useful?
12. Is water optimisation likely to be cost-effective, both now and in the future?

6.4 *Financial Analysis*

In considering the last point on the Check-List, cost benefits are likely to be both tangible and non-tangible.

Tangible benefits can be examined using appraisal techniques such as "Net Present Value" in order to compare capital and revenue items together.

Non-tangible benefits could include the avoidance of punitive fines and the loss of "environmental image". These should not be ignored in cost analysis, however difficult to quantify.

7 CONCLUSIONS

Water is no longer a cheap commodity and external regulatory demands are increasing significantly.

It may be time to give serious consideration to the optimisation of water.

Water recycling will make increasingly sound financial sense.

The cost of complex appraisal should be worthwhile.

8 ACKNOWLEDGEMENTS

Comments and advice from Diana Clark (Associate, WQM) have been much appreciated.

Why Recycle?, Rainbow (ed.) © 1994 Balkema, Rotterdam, ISBN 90 5410 367 1

The future is integrated waste management

D. Sherlock-Thomas
Associated Energy Projects, London, UK

ABSTRACT: The UK must recognise the importance of energy recycling as part of an integrated waste management strategy. Material recycling is a progressive process and expensive. Existing schemes are unlikely to offer a cost-effective way forward. Integrated waste management should be supported through national and local plans within which energy recycling should play a key role.

1. THE SCOPE OF THE PROBLEM

The municipal waste industry in Britain handles around 34 million tonnes of rubbish each year, of which the bulk - 92% - has historically gone to landfill sites around the UK. However, during the past few years - as the concept of recycling began to impact on public awareness - it has become widely accepted that the recycling of certain waste components (such as plastic, metal, glass, paper and cardboard) can offer significant environmental and economic benefits.

The typical percentage breakdown of British waste is:

Plastic	7%
Metal	8%
Glass	10%
Fines	10%
Misc textiles	12%
Putrescibles	20%
Paper & cardboard	33%

The industry is currently dominated by landfill as the disposal option with 92% of UK waste being disposed of by this method. The Government has set targets for recycling which it is hoped will result in 25% of municipal waste being recycled by the year 2000. In addition, the Government has requested that the packaging industry present planned recovery and recycling of 50%-75% of packaging by the year 2000.

Even if the current recycling target of 25% is achieved, local authorities will still have to dispose of the remaining, i.e 25.5m tonnes per annum.

2. WHAT IS INTEGRATED WASTE MANAGEMENT?

At the centre of the concept of integrated waste management is the ability to recover optimum value from the waste stream, so minimising costs, reducing volumes for landfill disposal and protecting the environment.

It is a mix of components starting with the actual collection of waste from homes and business premises. Collection systems appropriate to each community are developed and once collected the waste is treated in a number of different ways. Details of collection and treatment methods follow.

3. ELEMENTS OF INTEGRATED WASTE MANAGEMENT

3.1 Collection

Collection of waste can be achieved in a number of ways. First, by the traditional method of a door-to-door collection service. Secondly, by a door-to-door collection using the two bin system. With this system, waste is pre-sorted by the householder, separating specific materials such as paper, glass, plastics, cartons, and organic refuse, before collection. Both these systems are called 'Kerbside'.

A third method used is drive-in centres where waste segregation occurs in bulk. Finally, collection can be done through local banks which cater for glass, paper, aluminium and other materials. These latter two methods are known as 'Bring' systems. Once collected, the waste is transported, via transfer stations if need be, to treatment units whose functions are described briefly below.

Associated Energy Projects has long been aware of the direct link between collection and disposal. The way in which collection takes place to an extent dictates the quality of the material and therefore the types of processes to be used and therefore the quality of the recycled material. The artificial barriers set up in the UK through the different legislative arrangements and contract standards make it difficult for waste management companies to deliver integrated solutions.

Waste collection rests with district councils under the 1980 and 1988 Local Government Acts using compulsory competitive tendering with its emphasis on accepting the lowest bid. But waste disposal is the responsibility of county councils, metropolitan authorities and London boroughs under the Environmental Protection Act 1990 which can take environmental impacts into consideration when selecting winning bids from tenders.

As long as the dichotomy between collection and disposal is maintained it is very difficult for any coherent waste strategy to be agreed and implemented. There is a need for unitary Waste Authorities, whatever the results of the discussions taking place on local government reform.

3.2 Sorting and Recycling

On collection, waste is transferred to a sorting or recycling centre. Waste which has already been partially sorted at its origin through a 2 bin system can be much more easily and cheaply processed for recycling and re-use.

There are fundamentally two types of sorting/recycling plant. The first concentrate on mechanical systems and are capital intensive. The machinery is expensive, needs careful monitoring and requires regular maintenance. The second type of plant utilises labour to sift and sort materials. This provides for a very flexible system with staff capable of transfer from area to area to meet demand. However, this is demanding work for people and high health and safety standards are vital.

3.3 Digestion

a) Aerobic (Composting)

Some 3 to 4 million tonnes of putrescible household waste and 3 million tonnes of paper waste unsuitable for material recycling, together with another 2 million tonnes of garden waste collected at civic amenity sites, are consigned to landfills every year. Compost will, nevertheless, be of such a quality as to achieve a high enough retail price only when made of a high proportion of garden waste, also known as green waste.

In the UK composting is still regarded as experimental, Test Valley District Council achieve a recycling rate of 42% with composting accounting for 29%. As with all recycling results, the socio-economic make-up of the area together with the amount of resources committed to explaining Council policy may explain some trends.

In parts of continental Europe and the USA, where one plant in Baltimore can handle 700 tonnes a day, composting has gained a larger share of the market. At the scale of a particular local authority, green waste composting will lead to the development of small units in the 10-20,000 tonne a year range.

b) Anaerobic

The WARMER bulletin of May 1993 reported, "a full-scale 20,000 tonnes per year dry anaerobic digestion plant is to be built for the city of Salzburg (Austria) to process source separated organic wastes. Contained anaerobic digestion of remaining wastes enables all of the gas coming from the rotting material to be captured and used for electricity production. This follows the construction of a 10,000 tonnes per year plant in Brecht, Belgium".

The benefits of the Non Fossil Fuel Obligation (NFFO) can, of course be applied to this form of generation as any other. Anaerobic digestion is, however, still a relatively unusual form of reclaiming waste product and is limited to small sizes. There are many problems with separating out the organic wastes and the process can be very expensive.

AEP believes that this method may well be more appropriate for the use of sewage rather than municipal waste as the application of energy recycling can release the content of the waste more easily. Moreover, the end product which is comparable in weight to the input product, has to be disposed of by either landfill or energy recycling anyway, as it is not really a saleable product.

3.4 Landfill

Whilst the waste management field is at present tilted heavily in favour of landfill, it is important to recognise that differences of approach cannot be classed as a struggle between advocates of landfill and incineration.

In any waste management strategy there will always be a place and a need for landfill, particularly for inert waste, waste-from-waste and untreatable waste. Indeed, energy recycling itself produces waste which is landfilled, although there are now new processes to turn this into building blocks and aggregate materials.

The landfill industry is changing. Engineered landfill is now necessary and further new, stringent regulations are on the way from Europe. Costs will therefore rise steeply but landfilling raw waste still remains the only alternative or the cheapest option for some waste authorities. In conclusion, there will always be a need for landfill, but this capacity must be preserved.

Energy recycling tends to reduce the waste stream by 90% by volume, the key component of landfill, and 70% by weight, thus assisting the preservation process.

Preserving landfill in this way will ease public pressure against this form of waste disposal. Such pressure is having and is likely to have a continuing effect on the acceptability of landfill and therefore its availability, leading through to increases in cost per tonne. In many counties across the nation strong *"people power"* is halting the further development of large scale landfill sites.

3.5 Waste-to-Energy

For the United Kingdom the most overriding advantage of waste-to-energy is that it dramatically reduces the volumes of waste material destined for landfill. Existing landfill sites can therefore be conserved for much longer and the difficult task of gaining permission for new sites eased.

While public perceptions move against landfill the new generation of waste-to-energy plants exemplified by the newly built SELCHP (South East London Combined Heat & Power Ltd.) plant at Deptford offer the latest, proven technologies ensuring the highest environmental standards for the waste process.

This offers tremendous opportunities for the UK. The contribution that such plants can make to reducing reliance on fossil fuels for energy has already been recognised by Government, as it has in all advanced countries of the world.

What is needed is the development of a network of waste-to-energy facilities sensitively sized, appropriately planned and sited to best meet the needs of local communities.

3.6 Waste as a resource

One of the principles of integrated waste management is to recognise waste as a resource rather than a problem. Each year the UK produces wastes with an energy content equivalent to around 25 to 30 million tonnes of coal. Yet in 1993, Britain only generated around 2% of all its electricity from renewable sources.

This contrasts sharply with other developed countries, which utilise higher percentages of waste to produce energy:

Switzerland	80%
Japan	70%
Sweden	50%
Netherlands	45%
France	40%
Germany	40%
USA	20%
UK	7%

4. AURORA, AN INTEGRATED WASTE MANAGEMENT SOLUTION

4.1 What is Aurora

To help solve these complex problems AEP developed Concept Aurora. This represents a totally integrated solution, offering a complete package covering all aspects of solid-waste management from collection, transportation and sorting through to conditioning, recycling, composting, waste-to-energy incineration (energy recycling) and landfill.

Aurora is completely flexible in that it has the capacity to adapt and develop with time. It can be tailored to different community needs as not every community will have the same waste needs or choose the same priorities. Aurora also has the potential to develop with the changing requirements of different communities. Aurora's flexibility also allows local authorities to choose either to have all or individual components of the service or to establish a core service and add on other services at a later date.

At the heart of Aurora is the ability to recover optimum value from the waste stream, using whatever technologies and processes that are necessary, so minimising costs, reducing volumes for landfill disposal and protecting the environment.

4.2 The advantages of Aurora

The prime advantages of concept Aurora are that it breaks down the barrier between collection and disposal to offer an integrated solution which is flexible and can offer appropriate solutions for different communities.

Through bulk sorting and handling Aurora can offer viable, marketable products ofgood quality.

Where appropriate, all suitable material that

cannot be reused, can be incinerated using state-of-the-art technology to produce energy in form of heat and electricity.

Concept Aurora significantly reduces volumes going to landfill sites by thorough and very systematic sorting processes. Landfill locations are selectively filled and then restored to conform to local landscapes.

The organised collection and exemplary process technology combine to produce high quality waste with optimum value. The end products of the system are:

- conditioned recyclable material such as glass, aluminium, metal, paper and cardboard
- marketable good quality compost
- energy in form of electricity and heat.

4.3 The European Experience

Given the different pattern of development of the waste management industry in continental Europe, together with a much more acute shortage of landfill sites, it is no surprise that the major components of concept Aurora are more in evidence there than in the UK. Energy recycling is an integral part of the French waste management plan with nearly 40% of waste being used in this way.

Even Germany, which has excluded, for the time being, energy recycling from its Packaging Waste Law, incinerates 40% of its municipal waste. In essence, the whole balance of waste management is different in the rest of Europe and this has led to greater experimentation and a wider spread of technologies and approaches.

Aurora is already taking practical shape at Cergy-Pontoise, metropolitan suburb of Paris, at Nantes and at Nancy.

At Cergy-Pontoise a two-bin collection system with civic amenity centres and local banks is being implemented. The treatment centre is self contained with recycling, compost and energy recycling units - its low profile and imaginative architecture styled to blend in with local landscapes. All operations take place under cover eliminating air pollution and offensive odours.

Recyclable materials such as aluminium cans, metals and glass are automatically further segregated. Special arrangements are made for larger items such as refrigerators, furniture, batteries and oily materials. All sorted and pre-sorted materials are treated in a conditioning unit prior to being processed into recycled products.

5. RECYCLING TARGETS

5.1 Official Targets

The Government White Paper "This Common Inheritance" established the "challenging target of recycling half our recyclable household waste by the end of the century. That is around 25% of all household waste."

5.2 Existing systems

As stated already in this paper, there are two principal methods for separating waste at source; the "bring" system and "curbside" collection.

An interim evaluation by Warren Spring Laboratories of the costs and benefits of these systems has concluded that bring systems alone are unlikely to meet the Government's recycling target. Waste diversion rates of only 4% to 8% have been recorded in trials.

The figures show that higher diversion rates of all materials, except glass bottles, are generally being achieved in curbside schemes, although at a higher cost. The

WSL study confirms that bring systems generate more income in urban than in rural areas.

The study comments that "an integrated approach is seen by some to be the best way to achieve higher recovery rates."

5.3 A sustainable industry is needed

Perhaps the most important outcome of the WSL study is that of costs. The authors warn "that existing cost structures are unlikely to persist". Many of the schemes currently in operation divert income to charities and receive support from the business community.

"If the Government sincerely wishes to achieve its targets the involvement of industry as a whole, producers of waste and of recycled waste, users of recycled material and engineers with vision will be required to take over the recycling process in order to create a sustainable industry."

5.4 Community Involvement

Whilst industry, Government (local and national) and non-governmental organisations all have a role to play, AEP firmly believes that local residents should be encouraged to take responsibility and accept as a mission the proper handling of their wastes.

Based on our group's own experiences around the world, AEP is convinced that the best way is to ensure that local communities further develop responsibilites for their own waste. AEP has a history of full discussion and disclosure to local communities. This goes beyond dealing with local authority representatives and seeks to secure the positive involvement and support of each community.

With the development of improved plant and very new high standards of emission levels, the local community can choose waste to energy as a more environmentally beneficial option without fears that were associated with the old generation of incinerators.

Nevertheless, all those involved in waste management need to make it clear to the community that recycling is not a 'soft option', however it is done. Community involvement starts in particular by getting the citizen to recognise that proper waste management and recycling cost money.

Why Recycle?, Rainbow (ed.) © 1994 Balkema, Rotterdam, ISBN 90 5410 367 1

Integrated waste management strategies and environmental management systems – Case study Farnborough College, Hampshire, UK

T.N. Pinder
Farnborough College, Hampshire, UK

ABSTRACT: Recycling should not be seen in isolation but only as part of a wider integrated waste strategy. On the 1st September 1992 Farnborough College initiated an environmental management system which incorporated an integrated waste management strategy. The key step to this integrated approach is the adoption of a clear and detailed environmental policy. Farnborough's policy contains a commitment to waste minimisation, re-use and recycling, as well as examining purchasing policy and the environmental sustainability of all materials used. The success of the scheme, due in part to training and education of staff and students has turned a major annual waste management bill, into a self financing and marginally profitable exercise.

1 FARNBOROUGH COLLEGE

Farnborough College has developed an international reputation for providing high quality further and higher education and training by carefully considering the needs of both industry and students. Its management has energetically pursued a dynamic policy of achievement in all areas including environmental management. The college is situated in North Hampshire and offers a wide range of courses and facilities for its current student population of over 11,000. The college also has strong links with local industry and commerce, and this support has enabled it to keep at the forefront of advances in environmental management.

Farnborough has been fortunate in its steps towards an improved environmental image as it possesses a strong environmental technology department providing BSc and MSc courses in environmental management aimed at producing environmental managers for industry as well as regulation. This has provided a technical backup for the organisation and eased the implementation of its environmental system

2. ENVIRONMENTAL MANAGEMENT SYSTEMS

Effective systems for managing the environment are a vital part of any modern companies operations. Companies who aspire to total quality recognise that tight control on systems and practices are vital if continual and gradual reduction in environmental impacts are to occur. Amongst the many benefits offered by installing environmental management systems, improved efficiency and financial savings are common. At Farnborough the implementation of an environmental management system has aided the environmental integration of all aspects of the colleges operations.

Companies can either adopt a formal environmental management systems such as BS 7750 and EC Eco-management and audit regulations, or a less demanding simpler system. Whichever approach is used there is a need to set up a progressive cycle of setting policies, evolving strategies, implementing plans, and measuring progress.

3. FARNBOROUGH POLICY

Farnborough prepared its environmental policy statement in November 1990 and implemented an integrated waste strategy in September 1992. The environmental mission statement declares that the college "will undertake to promote environmental considerations throughout all aspects of its operations and delivery" (Potter 1992). To do this the environmental policy contains two distinct yet interwoven components:

- Firstly the college has initiated a programme of improvements which seek to 'green' the internal management practices of the college. Examples of this include the environmental awareness of cleaning and maintenance activities and use and disposal of items such as paper.
- Secondly as the internal management practices are brought in order, the goal has been to 'green' the colleges product, or in other words, its educational courses. This will require the incorporation of environmental awareness into all courses, from art to engineering.

In the delivery of its environmental mission the college and its management continually seek to fulfil the following policies:

1. Incorporation of an environmental ethos into all academic and institutional practices.
2. Promotion of energy conservation and the re-use and recycling of all materials used in the College's processes.
3. Wherever appropriate, use should be made of environmentally sustainable products and materials.
4. Staff and students should be encouraged to have an environmental responsibility.
5. Regular environmental audits should be maintained and updated covering the college practices and the environmental content of all college courses. This should include a review of the capacity and expertise to deliver environmental components in all its courses.
6. Resources should be provided for staff support so environmental responsibility in all courses can be effectively delivered.
7. An integrated and environmentally sound ethic should be provided across the spectrum of the college curriculum.

If grouped together policy 1 can be seen to be concerned with education and training staff and students in the benefits of sound environmental management. Policies 2,3 and 4 are more concerned with the colleges internal practices. And finally policies 5,6 and 7 are related to the 'greening' of the colleges product; its courses.

4. INTEGRATED WASTE MANAGEMENT

Environmental policies 2 and 3 were initially addressed by undertaking detailed waste audits of all materials used in the college. This 'mass-balance' approach, addressing all inputs and outputs from the campus, has been undertaken on a regular basis. Each item or material is then assessed in terms of:

- its environmental pollution potential and environmental sustainability;
- its re-use potential;
- its recycling potential.

This has required a detailed examination of the following key areas:

- all paper products including stationary, publicity material, toilet rolls, paper towels, etc;
- all caretaking materials and equipment including soaps, washing powders, polishes, etc;
- all maintenance materials and equipment such as paints, sealants, wood, metal, etc;
- all material and equipment involved in maintaining and landscaping the college grounds;
- all catering products and equipment, wrappings and containers such as cans, glass and plastic;
- all teaching materials including equipment, chemicals, solvents etc.

In order to determine the environmental suitability of each

item used in the College the following questions are asked:

- What is the possibility of utilising alternative materials and products that are more environmentally sustainable, or more readily re-used or recycled?
- What level of environmental pollution could result from its use?
- What is the energy demand involved in the use of specific materials and equipment?

Following this review process a comprehensive minimisation, re-use and recycling strategy has been adopted. This has involved a number of initiatives such as composting organic wastes, placing kitchen waste in wormeries, segregation of paper into several grades in order to obtain better prices, and collection of glass, drinks cans (aluminium and steel), ferrous and non-ferrous scrap for sale to traders.

5. IMPLEMENTATION

Successful implementation of an environmental management system is dependant to a large extent on the attitude of staff and students. Training for managers as well as staff was seen as vital if the scheme was to prove successful. Staff and managers have had access to environmental awareness training in order that they can incorporate environmental management into their courses and educate students in the benefits of good practice. Publicity and policing of the scheme around the college has also increased the success and return from the recycling scheme. Students are also beginning to benefit from receiving environmental awareness training as part of all their courses.

Financially the scheme in its first year broke even. The costs of collection of general waste which previously was under contract with the District Council has been covered by the proceeds of the recycling sales. Above all however the scheme is not intended to generate a profit, it is aimed at demonstrating that the college cares about the environment.

REFERENCES

Potter.J.F (1992) An Environmental Ethos for Educators and Trainers. *The Environmentalist* 12(2): 81-83

Why Recycle?, Rainbow (ed.) © 1994 Balkema, Rotterdam, ISBN 90 5410 367 1

Materials recycling and energy recovery – Partners not competitors

Kit Strange
Warmer Campaign, Tonbridge, UK

ABSTRACT: Global sustainable development is a justifiable environmental goal and is rapidly becoming a key factor in forming waste management practices and policies around the world. Presently far more attention is given to material recycling than to energy recovery, often without sufficient evaluation of the relative environmentalimpacts. There are good reasons to support a balanced approach to resource recovery. An integrated philosophy of harnessing energy and material will have significant environmental and economic benefits.

The 1992 Earth Summit in Rio produced 'Agenda 21' a blueprint for sustainable development, which calls on governments to adopt national strategies for sustainable development. Agenda 21 addressed the management of solid wastes, demanding national plans for waste prevention, re-use, recycling and safe treatment. Additionally, industrialised nations must plan to cap or reduce arisings by 2000. Governments should provide incentives, according to Agenda 21, for recycling and the recovery of energy from wastes. They should also establish guidelines for the safe re-use of waste and should encourage markets for recycled and re-used products.

Processes create wastes in proportion to the efficiencies of the conversion systems involved. Even sustainable development must accept the inevitable production of wastes. Society must simply aim to minimise the net environmental impact of the wastes it produces, through sound practices of waste management and resource recovery.

Many national and international guidelines and laws set out policies to ensure that such sound practices are implemented.

The European Commission defined EC environmental policy with the adoption, at the end of 1992, of its fifth environmental action programme "Towards Sustainability" (COM (92)23). This programme identified the management of wastes as a key task for the 1990s, describing targets for the management of municipal solid wastes (MSW).

These targets include, by the year 2000:

. stabilisation of quantities of MSW arising at EC average levels of 300 kgs per capita per annum
. recycling/re-use of paper, glass and plastics of at least 50 per cent (EC average)
. preparation of national waste management plans
. no export beyond EC borders for final disposal
. development of markets for recycled materials
. EC wide infrastructure for collection, separation and disposal

To achieve these targets, the European Commission envisages the following staged actions:

. introduction of landfill and packaging directives
. legislation and voluntary agreements to cease land fill for certain wastes
. generation of reliable data on MSW arisings, collec tion and disposal
. development of cleaner technologies and product design
. introduction of economic instruments
. implementation of MSW incineration emission stand ards for dioxins

Britain's own waste management policy mirrors the EC philosophy of shifting attention from disposal to prevention, minimisation and re-use. The 1990 Environmental Protection Act addresses waste minimisation in industry and also imposes tough constraints on national and local waste management practices.

Any formulation of policy must start by defining objectives, which is not a simple issue. Historically in Britain, and currently in the developing world, waste management served the needs of public health protection. Nowadays, policy development is informed by requirements for environmental protection, resource recovery, landfill preservation and market potential for recycled products.

While public health issues remain relevant to the debate on waste management, society has moved beyond such immediate demands. Within recent decades our population has grown in size and prosperity, demanding higher standards of resource recovery and cleaner industrial processes. These developments have been accompanied by a growing reluctance to accept locally-sited projects for waste management and resource recovery.

Society's waste management strategy is enshrined in what is known as the 'hierarchy of waste management', which seeks to stimulate waste avoidance or reduction, material re-use, recycling, energy recovery and final disposal.

HIERARCHY OF WASTE MANAGEMENT

The Royal Commission on Environmental Pollution (1993) has clearly restated this philosophy, with its declaration that waste management must be based on a four-stage decision procedure:

. wherever possible, avoid creating wastes
. where wastes are unavoidable, recycle them, if possible
. where wastes cannot be recycled in the form of materials, recover energy from them
. when the foregoing options have been exhausted, utilise the best practicable environmental option (BPEO) to dispose of the wastes

Different elements of the hierarchy are more amenable to action by different groups, and it is clear that the public in general is particularly enthusiastic about materials recycling. One reason for this is that the first stage in the process of recycling - separating out components from the waste stream - is more open to widespread public involvement.

The House of Lords Committee on the European Communities (1993) challenged the notion of a strict hierarchy, declaring that the belief that recycling is morally right is simplistic and environmentally unsound.

Within UK policy on waste management is the stated 'aspirational goal' to recycle 25 per cent of MSW by the year 2000. This target includes composting, but not waste avoidance, re-use or energy recovery.

The 25 per cent goal is arbitrary, based simply on the expectation that it should be achievable. It is not a diversion quota that stems from deep thought on what makes environmental or economic sense. Britain currently only achieves a material recycling rate of around 5 per cent, so any target probably represents a step in the right direction towards a coherent formulation of environmental policy.

Additionally, the British Government challenged the UK packaging industry to come up with a voluntary scheme to recover between 50 and 75 per cent, against the threat of legislation.

It seems unfortunate that as Britain moves towards policies on material recycling, she seems to find it hard to do the same with the recovery of energy from waste.

The next sections of this paper consider the environmental impacts of materials and energy reclamation.

ENVIRONMENTAL IMPACT OF MATERIAL RECYCLING

The recycling loop consists of four separate processes:

. collection and transport
. separation and cleaning
. reprocessing into new products
. marketing the new products

Each phase has its own environmental impact, which will differ according to material type, product and location.

Warren Spring Laboratory (Ogilvie 1992) studied resource use in recycling, by comparing the potential material and energy needs in making recycled products, against products made from virgin materials. The following conclusions were drawn for different materials:

Glass

Use of glass cullet offers energy savings up to 25 per cent. Solid wastes from recycling are around 5 per cent of cullet recycled, representing a potential saving of 95 per cent in solid wastes for disposal. The environmental impact of glass container re-use can be less than recycling, but benefits are dependent on trippage.

Plastics

Plastics have a high energy content (up to 40 MJ/Kg) compared to the energy demand of the recycling process (up to 15 MJ/Kg). Material recycling can also yield reduced emissions and solid wastes, although energy recovery may be the most appropriate option on occasions.

Paper and board

Recycling paper can save up to 70 per cent of energy needs. However, Britain is a net importer of pulp and paper produced using renewable energy sources. Paper production will always need new pulp sources, as recycling degrades paper quality.

Ferrous metals

Ferrous metals are easy to recycle, offering more than 70 per cent energy savings, 30 per cent less emissions to air and up to 70 per cent reduction in discharges to water. For each tonne of ferrous tinplate recycled, solid wastes are cut by 1.28 tonnes.

Non-ferrous metals

Aluminium is the main non-ferrous metal constituent in MSW where recycling brings environmental ben efits. Ninety-five per cent energy savings and 96 per cent reduction in air emissions have been claimed.

Organic fraction

Composting and anaerobic digestion are potential waste management options with resource recovery applications. Composting reduces waste volume and mass, diminishes the water content, produces a stable product and inactivates pathogenic organisms. Anaerobic digestion allows efficient energy recovery from biogas, affords more precise process control and less land space requirements than composting. It is inappropriate to compare these processes in terms of their environmental impact relative to unrecycled alternatives.

Textiles

Recycling textiles saves around half the energy de mand and reduces sources of pollution, particularly to water. As with paper recycling, textile reprocess ing degrades the fibre characteristics, so recycling cannot continue in perpetuity.

ENVIRONMENTAL IMPACT OF ENERGY RECOVERY

Incineration of MSW leaves a significant solid residue, whilst achieving a volume reduction of up to 90 per cent. The main sources of wastes from an incinerator are:

- Stack gases. Emissions to air which are not removed by the air pollution control systems .
- Liquid effluents. Water discharged from the pollu tion control equipment.
- Bottom ash. Contains inert slag and cinders, to gether with most residual metals.
- Fly ash. This stream is composed of fine particles within the exhaust gases, some metals and organics and products of incomplete combustion.

The pollutants arising from incineration fall into the following groups:

- Particulates. Fine particles of inorganic material with metals and adherent organic materials.
- Organic materials. Present as volatile compounds (including the class of dioxins), particularly as a result of incomplete combustion, in addition to particulates.
- Acid gases. Oxides of nitrogen and sulphur, hydro gen chloride and hydrogen fluoride.
- Metals. The group includes lead, cadmium, mer cury, arsenic, chromium, cobalt, copper, tin, manga nese, nickel and thallium.

The precise levels of pollutants will vary according the composition of the waste stream, conditions within the incinerator and the pollution control equipment installed.

Strict limits on emission from MSW in incinerators are demanded by EC legislation, including the MSW incineration directives (89/369/EEC and 84/429/EEC) introduced in 1989.

The Royal Commission on Environmental Pollution (1993) concluded that for most pollutants existing incineration plants in Britain produce small or negligible proportions of total UK emissions. The Royal Commission added that, even with very large increases in the amount of waste incinerated, the total pollution load from plants complying with new standards would not be a cause for concern - with the exception of mercury and cadmium.

The Commission noted that the new HMIP standards for emissions to air from incineration plants are more stringent (in some cases far more so) than for conventional combustion processes - which may present a danger of misallocation of resources.

Other forms of environmental impact from waste incineration plant will include: visual intrusion, noise, vehicle movements, local ecological effects. These impacts will be broadly equivalent to many other waste management or resource recovery infrastructural developments.

So, it is clear that materials recycling and energy recovery are both processes with their own potential for positive and negative environmental effects. Those effects need to be assessed in order to arrive at sensible, justifiable decisions on whether one resource option is preferable to another. The developing technique of Life Cycle Analysis (LCA) shows promise as a tool to assist decision making on environmental grounds.

LIFE CYCLE ANALYSIS

Life cycle analysis involves carrying out an inventory of the main impacts associated with the manufacture, use and disposal of a product, from the mining of the raw materials and energy used in its production and distribution, through to its use, possible re-use or recycling, and eventual disposal. A study would normally ignore second generation impacts, such as the energy required to fire the bricks needed to build the kilns used to manufacture the raw material.

Most LCAs are simply inventories of energy and material consumption and released to the environment: emissions to air and discharges to water and the level of solid waste generated. Carrying out an LCA is a relatively straight-forward exercise, providing the boundary of the study has been clearly defined, the methodology is rigorously applied, and the data is accessible.

Many companies have attempted to use data derived from full or partial life cycle analysis to promote their products. There has been argument about the relative merits of using aluminium versus steel cans, and paper cartons versus plastic or glass bottles, and indeed the environmental implications of using fresh as opposed to packaged frozen food.

Two of the main LCA controversies have been the battle between Scott Paper and Friends of the Earth UK over claims about the reduced impact on global warming of using virgin (as opposed to recycled) fibres in tissues, and the dispute over disposable versus non-disposable nappies. A study by Berlin University, commissioned by Procter & Gamble, came out in favour of disposables. It was immediately challenged by the Women's Environmental Network in London which commissioned its own study. Nappy studies have also been produced in the US by Arthur D Little, Franklin Associates, and others.

Studies on milk containers have been published in at least five European countries, most coming to different conclusions about which is environmentally more benign, the refillable glass or polycarbonate bottle, or the laminated paper carton.

A controversial German study in 1988 (published by the Federal Office of the Environment in Berlin) found plastic bags to be more environmentally benign than paper ones - studies in the US have come to similar conclusions about polystyrene as opposed to paper drinks cups.

A report published last year in the US, written by senior lawyers, concluded that results of LCA should not be used to advertise or promote specific products "until uniform methods for conducting such assessments are developed and a consensus is reached among government, business, environmental and consumer groups on how this type of environmental comparison can be advertised non-deceptively".

In September 1993 the results were published of a study conducted on behalf of the German Environment Ministry by the Fraunhofer Institute in Munich; the Institute for Energy & Environmental Research in Heidelberg; and the Packaging Research Company in Wiesbaden. The three and a half year study compared life cycles of four different forms of packaging for milk - returnable glass bottles, plastic pouches, and rectangular and gable-top paper cartons - considering among other details their use of raw materials, their energy demand and the amount of waste they generate. The study suggests that no overall conclusion can be drawn, and that the optimum form of packaging for milk depends on specific factors like distances transported, number of times refillables are actually refilled, and so on.

ECONOMIC COSTS, ENVIRONMENTAL BENEFITS

In theory, therefore, decision-making ought to be straightforward. With LCAs to provide relative environmental impact, one should be able to make choices as a producer, a customer, a waste disposal authority or a civil servant. One could even dream of rationalising decisions in terms of economic costs and environmental benefits, as illustrated in Figure 1. Thus through the influence of diminishing returns and economies of scale one could achieve optimal levels of reduction, re-use, recycling, energy recovery and final disposal. Each optimal level will represent the level of environmental protection where any amount spent or withheld would result in a reduced marginal environmental benefit. Of course any optimal position is only true for a particular product or material in a particular place at a particular time. In reality one can only hope to approach the ideal solution through time and experience, modifying and evolving policy and practice. Society achieves this through a tortuous, often random blend of regulation and economic instruments where the market is an insufficient economic driver.

As Barton (1991) has observed "it is one thing to have the desire to see an objective reached and quite another to obtain agreement on what is an acceptable cost and who should payit. The notion that to recycle materials may actually cost the householder...more than current disposal has not been fully taken on board by the general public".

So the development of international, national and local goals and policies tends to be patchy. Targets for recycling are commonplace and variable, targets for minimisation and energy recovery less common but equally disparate.

In America, nearly every state imposes a range of targets and incentives to increase recycling rates. Nine states have set minimum recycled content requirements for products and packaging. In other states voluntary agreements have averted legislation. Four states have imposed a moratorium on landfills and incinerators to force the pace of material reclamation. n Europe the situation is no clearer. The EC draft directive on packaging and packaging waste started out calling for 90 per cent recovery and 60 per cent recycling within then years. This was later diluted to 60 and 40 per cent, respectively. Later still there was talk of a 50 per cent recovery target. Almost every country in Europe has different targets and different timescales for different materials in MSW.

Environmental Costs/Benefits of Resource Management

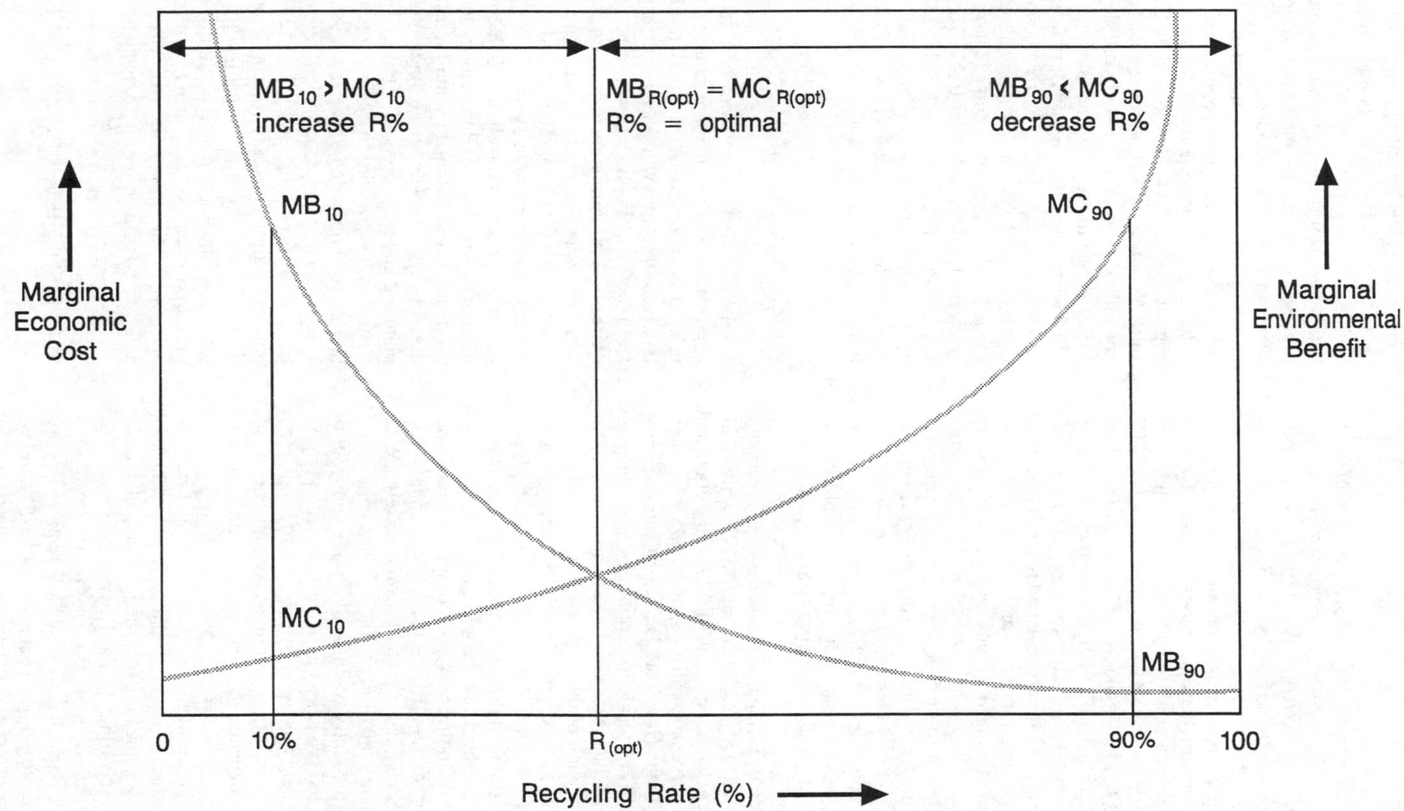

MATERIALS AND ENERGY - PARTNERS NOT RIVALS

There are many sources of evidence which support the contention that integrated resource recovery schemes bring benefits to the economy and the environment.

A survey by the Integrated Waste Services Association (IWSA 1993) studied the waste management practices in 70 American communities, reporting a close relationship between raised levels of energy recovery and materials recycling. The study concluded that communities with waste-to-energy (WTE) plants tend to have higher recycling rates than cities without such plants. The US Environment Protection Agency estimates that Americans recycle 17 per cent of their waste. The survey revealed recycling levels as high as 47 per cent for communities with WTE facilities. By combining materials and energy recovery, several states reduced the amounts of MSW 'exported' to other states, by more than 80 per cent. The mayor of Newark, New Jersey, attributed this success to the fact that communities which build advanced combustion plants also provide the infrastructure for effective recycling programmes, including household collection, separating and processing facilities, and public information.

New York City Council has agreed to develop a WTE plant (daily capacity 3,000 tonnes MSW), while planning to recycle 42 per cent of MSW by 2000 and achieve a 9 per cent overall reduction in MSW arisings. According to the New York Building Congress some 76 per cent of city residents supported this strategy, an enthusiasm which extended even to areas near to the proposed location of the WTE facility.

According to a recent public opinion survey in Austria, (Der Standaard, April, 1993), public support exists for combining material recycling with energy recovery. Of the 1,000 people polled, 65 per cent favoured incineration of waste, compared with 31 per cent supporting landfill disposal. Surprisingly perhaps, 53 per cent declared they would accept a WTE facility in close proximity to their own town.

Sweden - a country traditionally at the forefront of environmental protection policies - has opted to give priority to energy recovery from waste, to supplement its material recycling programmes. Already one household in ten depends on EFW as its main source of heating.

In Japan (Tanaka 1992) one sees a similar drive to increase resource recovery, partly due to a lack of indigenous resources, partly due to lack of landfill capacity. According to the Royal Commission on Environmental Pollution, Japan has nearly 2,000 MSW incinerators, virtually all of which recover energy. Japan incinerates some 72 per cent of all MSW and yet manages to achieve a materials recovery level of around 23 per cent.

There are numerous examples where plans are in hand to develop integrated resource recovery schemes. For example, Puerto Rico expects to develop a 1200 tonne per day WTE facility which will burn MSW after recyclables (ferrous metals and aluminium with an option for paper and plastics) have been recovered.

The relationship between materials recycling and energy recovery can be more than simply compatible, they can reinforce one another. Under most feasible recycling scenarios, the extraction of commercially useful streams of metals, paper, plastics and organics enhanced the fuel quality of the remaining fraction. Figure 2, based on data produced by ETSU for Hampshire County Council, demonstrates that the energy content of the residue rises with the removal from MSW of useful recyclables.

In Switzerland, the city of Zurich found economic benefits of integrating resource recovery options. Incineration of MSW in Zurich costs around SFr200/tonne, which generates sufficient revenue to fund waste disposal costs (Mäder 1993). Residents voted to spend SFr114 million on a WTE plant in the city centre, while material recovery is set to reach 40 per cent and the total amounts of MSW created per capita have declined steadily since the 1980s.

PUBLIC ACCEPTABILITY

Ultimately what society does with its waste will depend on society's collective aspirations and willingness to bear the costs. It is essential therefore that public acceptance is won for waste management projects and practices. Public support or acquiescence is required if communities are to develop material recovery facilities, waste-to-energy plants, new landfill sites, source separation programmes and any other facet of the waste management system. The need for open debate and public consultation is increasingly recognised by those who need to manage the wastes we all produce. Local authorities and industry are now addressing these needs as an integrated part of their planning and development (Barratt 1993).

The Royal Commission on Environmental Pollution (1993) recommended that "the waste management industry should adopt a policy of openness in providing information to the public". The House of Lords Select Committee on the European Communities (1993) declared that "information campaigns should be run locally, nationally and at Community level to explain the purpose and likely effect of return and management systems. The hierarchy (of waste management) and...the cost...also require explanation".

SCOPE FOR RESOURCE RECOVERY IN BRITAIN

Many studies into recycling rates have indicated that it will not be easy to achieve a national average of 25 per cent diversion. Thus the waste disposal authorities

The Effect of Recycling on the Gross Calorific Value of Household Waste

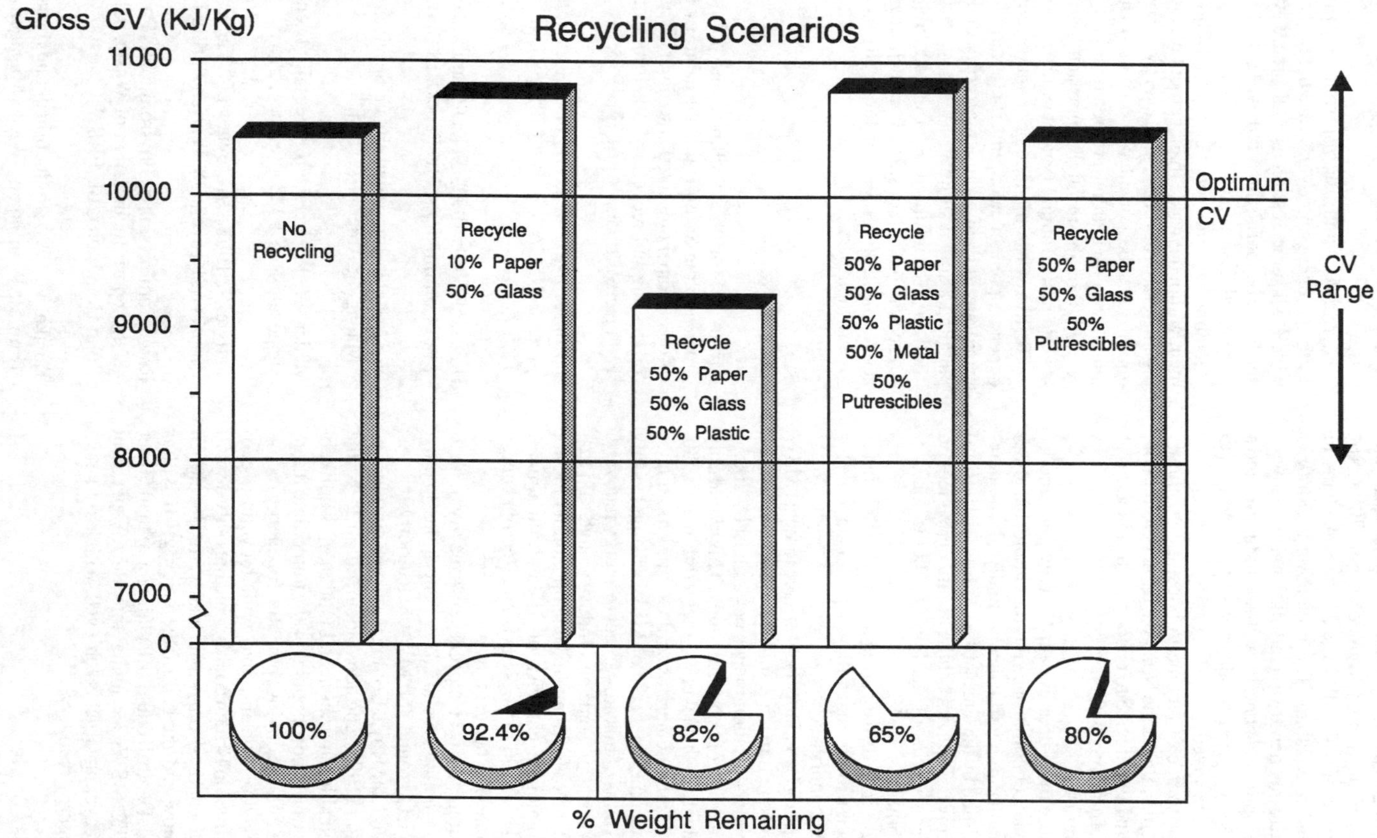

face a real challenge of what to do with the remainer. Warren Spring Laboratory (Barton 1991) have estimated that each household produces around 700 kgs of waste per year, of which 175 kgs can be considered to be clean and recyclable. The remaining majority is still a significant quantity to landfill. Energy recovery offers environmental benefits, volume reduction and a potential source of energy sales revenue, in addition to any gate-fee received.

One major source of income to boost the economic case for incineration is the Non Fossil Fuel Obligation, a market enabling device to encourage production of electricity from renewable energy sources such as domestic refuse.

In effect the main determinant for recycling/energy recovery is an economic one. It is inevitable that tough new EC regulations will cause virtually all incinerators to be closed by the end of 1996. A few will be retro fitted and any new plants will be highly dependent on the peculiar vagaries of the NFFO scheme for economic viability.

While there is landfill disposal to be had in Britain for a few pounds a tonne, that will be a preferred route for many local authorities.

Warren Spring Laboratory (1992) have forecast MSW incineration could reach 5 million tonnes per annum by 2001, and 5.4 million tonnes per annum by 2005 - around 15 per cent of present arisings. This would only be equivalent to the level of incineration in 1980. Britain currently incinerates some 2.5 million tonnes per annum (some 7 per cent of MSW arisings). Only around 0.8 mtpa of this are subject to energy recovery. The Department of Trade and Industry have said that they expect at least 7 plants to be in operation in 1995, with an aggregate capacity of 1.5 million tonnes, generating some 90 MW. This is an effective energy recovery rate of 5 per cent.

For the future, the DTI expects new capacity of up to 7 mtpa, consuming around 23 per cent of the 30 mtpa MSW to yield around 350 MW of electricity.

So, although the Royal Commission on Environmental Pollution (1993) considers incineration to be the Best Practicable Environmental Option, it is unlikely to become much more popular in the immediate future future. Despite the obvious potential for energy recovery to develop hand-in-glove with recycling, economic realities in Britain could mean that o nly a small handful of energy-from-waste plants are developed between now and the end of the century. It is thus possible that landfill could still claim more than 50 per cent of UK MSW, assuming that the 25 per cent material recycling target is achieved, with less than 25 per cent energy recovery.

CONCLUSIONS

Britain is set on a path towards sustainable development, in line with the objectives agreed in Rio at the 1992 Earth Summit. Any policy for sustainability will need to embrace the recovery and use of resources in waste.

The recovery and use of resources - material recycling and energy recovery - are themselves processes which consume resources and create wastes. This needs to be borne in mind when making resource recovery decisions.

Genuine resource recovery must embrac each element within the waste management hierarchy, including material recycling and energy recovery. Each option must be considered on its own merits and integrated into national, regional and local strategies. A well thought through approach will be able to operate to the mutual benefit of all.

Effective waste management and resource recovery depend upon an accurate appreciation of the true economic and environmental costs and benefits of pursuing a sound stratefy.

Full and free information is also a vital pre-requisite. All sectors of society should be encouraged to play an active role in this important debate, for everyone needs to play their part.

REFERENCES

Barratt, R.D. 1993. Technical and political issues related to the South Hampshire energy from waste project. Paper presented at seminar "Waste handling, processing and recycling". Institution of Mechanical Engineers, London April 27, 1993.

Barton, J. 1991. Warren Spring Laboratory. Paper given to NSCA, Lincoln College, Oxford. April 10/ 11, 1991.

House of Lords 1993. Select Committee on the Euro pean Communities, HL Paper 118 (24).

IWSA 1993. Integrated Waste Services Association. "Managing America's Trash".

Mäder, W. 1993. Environmentally friendly waste disposal. Rec '93 Conference. p96

Ogilvie, S. 1992. "A Review of : The Environmental Impact of Recycling", Warren Spring Laboratory.

Royal Commission on Environmental Pollution. 1993. 17th Report: Incineration of Waste. Chairman: Sir John Houghton CBE FRS

Tanaka, M. 1992. Reduction of and resource recovery from MSW in Japan. Waste Management and Re sources 10:453-459.

Why Recycle?, Rainbow (ed.) © 1994 Balkema, Rotterdam, ISBN 90 5410 367 1

Information for waste management – The Waste Management Information Bureau and its sources

Diana Silver
Waste Management Information Bureau, AEA Technology, Harwell, UK

ABSTRACT: The UK Waste Management Information Bureau (WMIB) is the national referral centre for advice and information on industrial and municipal waste management. It also provides an international service to any organisation or individual who needs to know about the subject. This paper outlines its services and discusses a selection of the most useful sources of information, particularly on recycling.

INTRODUCTION

Up-to-date, good quality information is of paramount importance to any organisation working in any field. However, it is not always easy to get speedy access to the most important information when there is a great deal of information available on a subject. This is particularly true in the case of environmental topics. It is therefore important to know where to look or who to ask for the most relevant sources of information.

The purpose of this paper is to outline the services of the Waste Management Information Bureau (WMIB) and explain how it can be of assistance in providing information and advice on all aspects of waste management, including recycling. Some of the main sources which are used by the Bureau to answer enquiries and maintain the WasteInfo database are also covered.

1 WASTE MANAGEMENT INFORMATION BUREAU (WMIB)[1]

1.1 *Background*

In the early 1970s there were a number of pollution incidents in the UK which brought to light the lack of a central information resource on hazardous waste management. The Waste Management Information Bureau (WMIB) was set up in 1973 with funding from the UK Department of the Environment. Harwell Laboratory in Oxfordshire was chosen as the location, given the expertise in the management of hazardous materials held within Harwell's Hazardous Materials Service (which later became the Environmental Safety Centre). The intention, from the beginning, was not solely to supply information, but also to provide advice on hazardous waste management to all enquirers (although at first most enquiries came from industry and local government). After a short time the coverage was expanded to cover all aspects of waste management (i.e. not only hazardous waste management) and enquiries started coming in from all sectors of the community.

1.2 *Subject Coverage*

WMIB now covers all aspects of non-nuclear, industrial, domestic and commercial, solid, sludge, liquid and gaseous waste management. A wide range of waste management issues are covered, including:

- waste minimisation
- recycling and recovery
- collection, treatment and disposal, and
- a large range of associated environmental topics.

Taking two of these areas in more detail, the coverage includes:

- Recycling/recovery
  - recycling of municipal solid waste, including collection issues, policy, economics and technologies

- energy from waste, including refuse derived fuel (RDF), and energy recovery from incineration and landfill gas
- separation technologies
- reclamation of land, including the clean-up of contaminated land and landfill restoration.

- Treatment and disposal
  - physical, chemical and biological treatment of waste
  - all aspects of landfill from planning, design and licensing to post-closure, including research into technologies for management of leachate and landfill gas, policy issues and economics.
  - incineration - all aspects including siting and design, technologies, emission control, policy issues and economics
  - Other existing technologies or technologies under development.

Political and economic issues influencing waste management are becoming increasingly important and therefore WMIB is now covering these in more detail. For instance, we hold documents which describe the effectiveness of existing legislation and its enforcement, the development of new policies and regulations, and the impact of international initiatives on the policies of national governments. The economics of different approaches to waste management is also becoming a regular pre-occupation of our enquirers!

To a lesser extent WMIB also covers information on companies which provide equipment or services in the waste management field, especially those which are innovative or operate at a large-scale.

Other more general environmental issues which have an impact on waste management are also covered to some extent, for example environmental auditing, environmental impact assessment, global warming, and ozone depletion.

2 SOURCES

The following section is an overview of the sources of information used by WMIB in answering enquiries. It is not a comprehensive list but does discuss a selection of the many information sources available.

2.1 *Databases*

2.1.1 *WasteInfo*[2]

From the time WMIB was originally set up, the literature which has been acquired to answer the enquiries has been compiled into a bibliographic database called WasteInfo.

WasteInfo is a bibliographic database i.e. each record contains information such as the author, title, the source of the document and an abstract. It now contains over 85,000 references and closely reflects the subject coverage of WMIB. Only published information sources appear on the database as it is intended to be a guide to the literature available to the user.

All the usual sources are covered, such as journals, books, reports, conferences, patents, theses, newspapers, press releases, legislation, company literature etc. The coverage is international and literature in all the major EC languages is covered. All abstracts are in English but the titles are in the original language, with an English translation. As there is so much published on the subject, only good quality and long lasting information is selected. The type of material ranges from the highly technical through to discussion papers.

WasteInfo is searched on behalf of enquirers by WMIB staff but it is also available online and on CD-ROM. A selection of the items which enter the database each month is also published in the Bibliographic section of WMIB's international journal WASTE & ENVIRONMENT TODAY[3].

2.1.2 *Other Online and CD-ROM Databases*

We do not restrict ourselves to our own database when answering enquiries, as WMIB has access to a very wide variety of sources including other online databases. As far as we know, there are no specific recycling bibliographic databases available online on the main online hosts, and no other specific waste management databases apart from our own. However, there are a number of more general environmental databases and databases which are specific to an industry. For instance, Enviroline[4] and Pollution Abstracts[5] are wide ranging environmental databases and useful to us as a backup to WasteInfo or for searching information on subjects which are on the periphery of our primary area of interest. Chemical Abstracts[6], Aluminium Industry Abstracts[7], RAPRA[8], PIRA[9], Metadex[10], and Aqualine[11] are amongst the industry-specific databases which can be of great help when looking for information on the recycling of a particular material or for a particular technology.

There are a number of other databases which are available by contacting the producer direct. There are a few directories available which will list these databases, one of which, for the UK, is the ECO Directory of Environmental Databases in the United Kingdom[12].

2.2 *Printed Sources*

2.2.1 *Journals*

Environmental publications of all types are on the increase, but none more so than journals and newsletters. There are a number of general environmental, waste management or specifically recycling journals which are, nevertheless, considered by WMIB to be of good quality. Specific recycling journals include: Warmer Bulletin[13] (UK and international), The Management of World Wastes[14] (US) and its international edition, Biocycle[15] (US), Resources Conservation and Recycling[16] (international), Resource Recycling[17] (US), Solid Waste and Power[18] (US) and Materials Reclamation Weekly[19] (UK).

Industry-specific journals are very good sources of information on the environmental impact of the particular industry's products and processes. The trade associations and industry councils often publish their own newsletters.

2.2.2 *Directories and Year Books*

There has been a dramatic increase in the number of directories available to the information scientist in the last couple of years. Two specific recycling directories for the UK are the Waste Watch National Directory of Recycling Information[20], and the UK Recycling Directory[21] by Recycling and Resource Management Magazine. A very useful directory on companies involved in recycling and reclamation in the UK is the Materials Reclamation Handbook[22], published by EMAP, Maclaren Ltd who also publish the Materials Reclamation Weekly. A similar directory for North America is an annual special edition of the journal, Resource Recycling[23]. For Europe there are two general environmental company directories which we use frequently - Polmark: European Pollution Control and Waste Management Industry Directory[24] and the ENTEC Directory of Environmental Technology[25], which is in English, German, French and Spanish.

There are several UK general environmental directories and year books which cover recycling, including Waste Recycling and Environmental Directory[26] (Thomas Telford) and Environment Industry Year Book[27] (the Environment Press). A general environmental directory for North America is World Environmental Directory: North America[28] (Business Publishers Inc).

One of our most frequently used publications is Croner's Waste Management[29], which is an updated, loose-leaf publication covering legal aspects, practical waste management information, waste charts and a useful directory which lists recycling companies. Its value is enhanced by the fact that it is amended regularly.

2.2.3 *Books, Reports and Conferences*

2.2.3.1 *Books*

I have picked a very small selection of books to list here. The ISWA Working Group on Recycling, established in 1989, is now publishing reports of its work. The first edition of Recycling in Member Countries[30] was published in 1992 and includes information on Denmark, Italy, Norway, UK, Spain and Sweden. Another European source is Solid Waste Recycling: Municipal Solid Waste Recycling in Western Europe to 1996[31] (Elsevier).

The GJ Press Inc who publish Biocycle have also published a guide called The Collecting, Processing and Marketing of Recyclables[32]. It covers only municipal solid waste but takes a thorough look at each step in turn. There is also an extensive report on materials recovery facilities (MRFs).

The Steel Can and Recycling Information Bureau is responsible for a useful publication called Recyclopedia: Recycling Domestic Waste - Meeting the Challenge[33]. Despite the fact that the publisher is obviously involved in a specific material, the book covers all aspects of MSW recycling and, although written from a UK perspective, there are useful comparisons with other countries. The book contains a number of useful statistics.

2.2.3.2 *Reports*

Warren Spring Laboratory[34] has published a number of excellent reports on a range of recycling issues and they produce a regular list of these reports.

Market research reports are very useful but are also very often extremely expensive. WMIB includes the details of many of these in WasteInfo by taking the information from publicity material, but there are a few which WMIB is willing to purchase. Mintel published a market intelligence report in October 1991 on recycling in the UK costing £195[35] (they also published a special report on recycling in 1992 costing £895). There are other more general waste management market reports which cover recycling. We think very highly of the Marketing Strategies for Industry (MSI) Data Report: Waste Management: UK[36]. The Centre for Exploitation of Science and Technology (CEST) has also published a market report by Aspinwall and Company on waste management, again covering the UK[37].

There are of course numerous organisations which produce reports on recycling, both general and technical. Trade associations are often prolific in publishing such reports. Reports funded by national governments are very good sources of information and their existence can usually be picked up from bibliographic databases or, if they are significant, they are often referred to in news journals.

2.2.3.3 *Conferences*

Every month seems to bring yet another annual conference on some aspect of the environment! Conference proceedings can often be good sources of information on recent technologies, although the same papers can often appear in a number of similar conferences. A good source of information about the existence of conferences is the diary of forthcoming events included in several relevant journals, including our own WASTE & ENVIRONMENT TODAY journal.

The International Solid Waste and Cleansing Association[38] (ISWA) holds their major International Solid Waste Congress every four years and also an annual conference. They also hold specialised conferences from time to time.

Waste Management International[39], published by EF-Verlag für Energie-und Umwelttechnik GMbH, provides a thorough exploration of recycling of both industrial and domestic wastes. Despite its title it concentrates on recycling (the previous editions were entitled Recycling International). It comes in the form of a collection of conference papers from the Internation Recycling Congress, but the conference details do not appear anywhere in the publications. The congress is held every 3 years and the 1992 publication comes in two volumes. Its coverage is international.

Other general environmental international conferences include, among others: Polmet[40], Elmia[41], and Pollutec[42]. Specific recycling conferences include: the International Conference on Solid Waste Management and Secondary Materials[43], the World Recycling Conference and Exhibition[44], the International Forum on Resource Recovery from Waste[45], the International Conference of the International Federation of Reclamation and Recycling Industries[46], and the International Symposium on Combustion and Energy Recovery from MSW[47]. There are many conferences on particular recycling issues such as the European Congress on Packaging and the Environment[48], or the Reclamation, Treatment and Utilisation of Coal Mining Wastes[49]. Many countries have very well respected national annual conferences. For instance, in the USA, there is the TAPPI Environmental Conference[50], the Madison Waste Conference[51] and the National Waste Processing Conference[52].

National trade associations and professional bodies often hold their own annual conferences.

2.2.4 *Official Publications*

Of course, official government publications are of paramount importance when investigating any subject. Apart from national legislation, it is important to keep up to date with official guidance documents, codes of practice and so forth. In the UK, for instance, the Department of the Environment[53] (DoE) produces consultation papers prior to issuing new or revised regulations and guidance documents. It is essential that interested parties have speedy knowledge of the publication of these consultation papers. The DoE issues press releases periodically to inform the public of the consultation papers. The publisher of official government publications in the UK is HMSO, which issues a Daily List[54] of new publications.

One of the most important series of guidance documents in the UK for waste management is the Waste Management Papers series. These are constantly reviewed and updated. Waste Management Paper No 28 covers recycling[55]. There are now several loose-leaf, updated manuals covering legislation. Nigel Haigh has compiled the Manual of Environmental Policy - the EC and Britain[56] (Longman). Meaningful statistics always seem difficult to get hold of. For the UK, the source of official published statistics on the environment is the Digest of Environmental Protection and Water Statistics[57] from the DoE. Recently the DoE has also published The UK Environment[58] - a "state of the environment" type of publication which many other national governments produce regularly. Eurostat regularly publishes Environment Statistics[59], in German, English and French, through the EC Official Publications Office, and the OECD produces a similar publication - OECD Environmental Data[60]. The United Nations and the EC collaborate in publishing The Environment in Europe and North America: Annotated Statistics[61].

2.2.5 *Organisations*

There are many trade associations, pressure groups and official bodies who may be contacted for information on recycling. They are also often a valuable source of printed information.

In particular, in the UK, there is Waste Watch[62] and the Warmer Campaign[63] who both encourage the recycling of materials and energy from waste. Most of the main recyclable materials are represented by industry-sponsored associations and councils. For example, Incpen is the Industry Council for Packaging and the Environment[64], the "Alliance" is the Alliance for Beverage Cartons and the Environment[65], and ICER is the Industry Council for Electronic and Electrical Equipment Recycling[66]. Then there is, of course, the Recycling Council[67].

Most countries have similar associations and many of these councils are involved in Europe-wide or international bodies. The waste management industry itself has its trade associations and professional institutes. In the UK these are: the National Association of Waste Disposal Contractors[68] (NAWDC), and the Institute of Wastes Management[69] (IWM) respectively. Pressure groups who have an interest in recycling include Friends of the Earth[70] and Greenpeace[71].

Then, of course, there is the Waste Management Information Bureau....

3 WMIB SERVICES

All of the sources outlined above, whether printed or not, are referred to by the staff of WMIB when answering enquiries. WMIB staff will answer short enquiries free of charge (approximately 10-15 minutes). A range of services can be offered to the fee-paying client, including searches of WasteInfo or of other online databases.

In conclusion, the Waste Management Information Bureau is available at the end of the phone or fax to assist anyone who needs help in pinpointing the most useful sources of information on a waste-related subject, or who needs answers to specific problems

REFERENCES

(1) Waste Management Information Bureau, AEA Technology, 7.12 Harwell, Didcot, Oxon, OX11 ORA, UK. Tel: +44 235 433442/432334; Fax: +44 235 432854.

(2) WasteInfo: Producer: Waste Management Information Bureau. Online host: InfoPro Technologies. CD-ROM - SilverPlatter Inc.

(3) WASTE & ENVIRONMENT TODAY, Waste Management Information Bureau. Tel: +44 235 433484. Monthly.

(4) Enviroline: Producer: Bowker-Saur. Online hosts include Dialog/Data Star, ESA-IRS, InfoPro Technologies, DIMDI, FIZ-Technik. CD-ROM - Bowker-Saur.

(5) Pollution Abstracts: Producer: Cambridge Scientific Abstracts. Online hosts include Dialog/Data Star, ESA-IRS, InfoPro Technologies, DIMDI. CD-ROM - PolTox from SilverPlatter Inc.

(6) Chemical Abstracts (CA Search): Producer: Chemical Abstracts Service. Online hosts include Dialog/Data Star, ESA-IRS, InfoPro Technologies, STN, DIMDI.

(7) Aluminium Industry Abstracts: Producer: Materials Information - ASM International/ Institute of Materials. Online hosts include Dialog, ESA-IRS.

(8) RAPRA Abstracts: Producer: RAPRA Technology Ltd. Online hosts include

Dialog/Data Star, InfoPro Technologies, ESA-IRS, STN. CD-ROM - RAPRA.

(9) PIRA Abstracts: Producer: PIRA. Online hosts include Dialog/Data Star, STN, InfoPro Technologies. CD-ROM - Dialog OnDisc.

(10) Metadex: Producer: Materials Information - ASM International/Institute of Materials. Online hosts include Dialog/Data Star, ESA-IRS, InfoPro Technologies, STN, FIZ-Technik. CD-ROM - Dialog OnDisc.

(11) Aqualine: Producer: WRc (Water Research Centre). Online host: InfoPro Technologies. CD-ROM - WRc.

(12) ECO Directory of Environmental Databases in the United Kingdom. Barlow, M., Fleming, P., Button, J. (eds). ECO Trust, 10-12 Picton St., Montpelier, Bristol BS6 5QA, UK. Tel: +44 272 420162. 1992. ISBN 1 874666 00 8. £35. New edition due out September 1994.

(13) Warmer Bulletin. The Warmer Campaign, Bridge House, High Street, Tonbridge, Kent TN9 1DP, UK. Tel: +44 732 368333. Quarterly.

(14) World Wastes, PO Box 41369, Nashville, TN 37204-1094, USA. Monthly. International Edition, Circulation Dept, 6151 Powers Ferry Road NW, Atlanta, GA 30339, USA. Tel: +1 404 955 2500. Bi-monthly.

(15) Biocycle. J G Press Inc, 419 State Ave, Emmaus, PA 18049, USA. Tel: +1 215 967 4135. Monthly.

(16) Resources, Conservation and Recycling. Elsevier Science Publishers BV, PO Box 1527, 1000 BM Amsterdam, The Netherlands. Tel: +31 20 5803258.

(17) Resource Recycling. 1206 NW 21st Ave, PO Box 10540, Portland, OR 97210, USA. Tel: +1 503 227 1319. Monthly.

(18) Solid Waste and Power. HCI Publications, 410 Archibald Street, Kansas City, MO 64111-3046, USA. Tel: +1 816 931 1311. Seven issues per year.

(19) Materials Reclamation Weekly. EMAP Maclaren Ltd, PO Box 109, Maclaren House, Scarbrook Road, Croydon, Surrey CR9 1QH, UK. Tel: +44 81 688 7788. Weekly.

(20) National Directory of Recycling Information, 3rd Edition. Waste Watch, see (62). 1993. ISBN 1 898026 01 7. £9.95.

(21) UK Recycling Directory 1992-93. Recycling and Resource Management Magazine, Oakwood House, 3 Moulton Park Office Village, Northampton NN3 1AP. Tel: +44 604 790707. 1992. £18. Bi-annual. 1994/95 edition due out 1994.

(22) Materials Reclamation Handbook 1993. EMAP Response Publishing, Wentworth House, Wentworth Street, Peterborough PE1 1DS, UK. Tel: +44 733 63100. 1993. ISBN 1 85438 180 6. £30. 1994 edition available February 1994. ISBN 1 85438 245 4 £30.

(23) Resource Recycling. See (17).

(24) Polmark: European Pollution Control and Waste Management Industry Directory 1993. ECOTEC Research and Consulting Ltd. Frost and Sullivan Ltd, Sullivan House, 4 Grosvenor Gardens, London, SW1W 0DH, UK. 1993. ISBN 0 86621 000 8. £255.

(25) ENTEC Directory of Environmental Technology. European Edition. The Entec Press. Kogan Page Ltd, 120 Pentonville Road, London, N1 9JN, UK. Tel: +44 71 278 0433. 1993. ISBN 0 7494 0853 7. £125.

(26) Waste Recycling and Environmental Directory 1993. Thomas Telford, Thomas Telford House, 1 Heron Quay, London, E14 4JD, UK. Tel: +44 71 987 6999. 1992. ISBN 0 7277 1845 2. 1994 edition due out end 1993. £40.

(27) Environment Industry Yearbook 1994. The Environment Press Ltd, 26 Brook Street, Bath BA1 2LN, UK. Tel: +44 225 330312. 1993. ISBN 0 9519096 0 3. £72 (£67 by credit card).

(28) World Environmental Directory: North America. 6th Edition. Business Publishers

Inc, Silver Spring, Maryland 20910-4464, USA. Tel: +1 301 587 6300. 1991. ISBN 0 916 742 06 7. $233. 7th Edition due out in 1994.

(29) Croner's Waste Management, Croner Publications Ltd, Croner House, London Road, Kingston-upon-Thames, Surrey, KT2 6SR, UK. Tel: +44 81 547 3333. 1993. ISBN 1 85524 193 5. £156.50 including one year updates. Subsequent years approximately £95.

(30) Recycling in Member Countries. ISWA Working Group on Recycling. See (38). 1992. Approximately £48.

(31) Solid Waste Recycling. Elsevier Advanced Technology, Mayfield House, 256 Banbury Road, Oxford, OX2 7DH, UK. 1992. ISBN 1 85617 138 8. £695.

(32) The Collecting, Processing and Marketing of Recyclables. J G Press Inc. See (15). 1990. ISBN 0 932424 11 2. $59.95.

(33) Recyclopedia: Recycling Domestic Waste - Meeting the Challenge. Steel Can Recycling Information Bureau, 69 Monmouth Street, London WC2H 9DG, UK. Tel: + 44 71 379 1306. 1993. ISBN 0 946555 28 1. £12.99.

(34) Warren Spring Laboratory, Gunnels Wood Road, Stevenage, Herts, SG1 2BX, UK. Tel: +44 438 741122.

(35) Recycling. Mintel International Group Ltd, 18-19 Long Lane, London EC1A 9HE, UK. Tel: +44 71 606 6000. 1991. £195.

(36) MSI Data Report: Waste Management: UK. Marketing Strategies for Industry (UK) Ltd. Viscount House, Riverside Business Park, River Lane, Saltney, Chester CH4 8QY, UK. Tel: +44 244 681424. 1993. £330.

(37) Waste Management. Aspinwall and Company. Centre for Exploitation of Science and Technology (CEST), 5 Berners Road, London N1 0PW, UK. Tel: +44 71 354 9942. 1990.

(38) International Solid Waste and Cleaning Association (ISWA), Bremerholm 1, DK 1069 Copenhagen K, Denmark. Tel: +45 33 91 44 91.

(39) Waste Management International. Thomé-Kozmiensky K J (ed). E F -Verlag fur Energie- und Umwelttechnik GmbH, Erich-Dieckhoff-Str. 50, 0-1950 Neuruppin, Germany. 1992. ISBN 3 924511 64 0. Approx £120. Next edition due 1995.

(40) Polmet - Pollution in the Metropolitan and Urban Environment. Hong Kong Institution of Engineers, 9/F Island Centre, 1 Great George Street, Causeway Bay, Hong Kong.

(41) ELMIA. Box 6066, S-550 06 Jönköping, Sweden. Tel: +46 36 15 20 00.

(42) Pollutec. Groupe SEPFI TECHNOEXPO, 8 rue de la Michodière, 75002 Paris, France. Tel: +33 1 47 42 92 56.

(43) International Conference on Solid Waste Management and Secondary Materials. Department of Civil Engineering, Widener University, 1 University Place, Chester, PA 19013-5792, USA. Tel: +1 215 499 4042.

(44) World Recycling Conference and Exhibition, Resource Recycling Magazine. See (17). Tel: +1 203 852 0500.

(45) International Forum on Resource Recovery from Waste. CISA, Via Marengo 34, I-09123 Cagliari, Italy. Tel: +39 70 271652.

(46) International Conference of the International Federation of the Reclamation and Recycling Industries (BIR). Rue du Lombard 24, Bte 14, B-1000 Brussels, Belgium. Tel: +32 2 514 2180.

(47) International Symposium on Combustion and Energy Recovery from MSW. Carbone Congressi, via Q Sella 68, 90139 Palermo, Italy. Tel: +39 91 321 273.

(48) European Congress on Packaging and the Environment. IBGE, Gulledelle 100, B-1200 Brussels, Belgium. Tel: +32 2 775 75 75.

(49) Reclamation, Treatment and Utilisation of Coal Mining Wastes. UK Recycling Council. See (67).

(50) TAPPI Environmental Conference. TAPPI, PO Box 105113, Technology Park, Atlanta, GA 30348-5113, USA. Tel: +1 404 446 1400.

(51) Madison Waste Conference, Department of Engineering Professional Development, University of Wisconsin - Madison, 432 N Lake Street, Madison, WI 53706, USA. Tel: +1 608 262 2061.

(52) National Waste Processing Conference, American Society of Mechanical Engineers, United Engineering Center, 345 East 47th Street, New York, NY 10017, USA.

(53) UK Department of the Environment (DoE). 2 Marsham Street, London SW1P 3EB, UK. Tel: +44 71 276 3000 (General address and telephone number).

(54) Daily List. Her Majesty's Stationary Office (HMSO), PO Box 276, London SW8 5DT, UK. Tel: +44 71 873 8499. Daily = £85, Weekly = £50.

(55) Waste Management Paper 28: Recycling. Department of the Environment. HMSO. See (54). Tel: +44 71 873 9090. 1991. ISBN 0 11 752445. £9.

(56) Manual of Environmental Policy - the EC and Britain. Haigh, N. Longman Group Ltd, Westgate House, The High, Harlow, Essex CM20 1YR, UK. Tel: +44 279 442601. 1992. ISBN 0 582 08715 5. £175. Updates at individual prices.

(57) Digest of Environmental Protection and Water Statistics, No 15, 1992. Department of the Environment. HMSO. See (54). 1993. ISBN 0 11 752815 3. £14.95.

(58) The UK Environment. Department of the Environment. Brown, A. (ed). HMSO. See (54). 1992. ISBN 0 11 752420 4. £14.95.

(59) Environment Statistics 1991. Eurostat. Office for Official Publications of the European Communities. L-2985 Luxembourg. 1992. ISBN 92 826 4666 1. ECU 11 (HMSO £9).

(60) OECD Environmental Data 1993. OECD, 2 rue André-Pascal, 75775 Paris 16, France. August 1993. ISBN 92 64 03512 5. (HMSO £42).

(61) The Environment in Europe and North America: Annotated Statistics 1992. United Nations, Economic Commission for Europe. United Nationals Sales Section, New York, USA or Geneva, Switzerland. ISBN 92 1 116537 7. (HMSO £35).

(62) Waste Watch, Hobart House, Grosvenor Place, London SW1X 7AE, UK. Tel: +44 71 245 9998.

(63) See (13).

(64) Industry Council for Packaging and the Environment (INCPEN). Tenterton House, 3 Tenterton Street, London W1R 9AH, UK. Tel +44 71 409 0949.

(65) Alliance for Beverage Cartons and the Environment. St Mary's Centre, Oystershell Lane, Newcastle-upon-Tyne, NE4 5QF, UK. Tel: +44 91 232 0225.

(66) Industry Council for Electronic and Electrical Equipment Recycling (ICER), 6 Bath Place, Rivington Street, London EC2A 3JE, UK. Tel: +44 71 729 4766.

(67) Recycling Council, Dunston Hall, Dunston, Stafford, Staffordshire ST18 9AB, UK. Tel: +44 785 712666.

(68) National Association of Waste Disposal Contractors (NAWDC), Mountbarrow House, 6-20 Elizabeth Street, London SW1W 9RB, UK. Tel: +44 71 824 8882.

(69) Institute of Wastes Management (IWM), 9 Saxon Court, St Peter's Gardens, Northampton NN1 1SX, UK. Tel: +44 604 20426.

(70) Friends of the Earth (FoE) Ltd, 26-28 Underwood Street, London N1 7JQ, UK. Tel: +44 71 490 1555.

(71) Greenpeace Ltd, Cannonbury Villas, London N1 2PN, UK. Tel: +44 71 354 5100.

Why Recycle?, Rainbow (ed.) © 1994 Balkema, Rotterdam, ISBN 90 5410 367 1

Alternative aggregates supply sources: The case for using recycled mineral and construction wastes

Andrew Marsay
Arup Economics & Planning, London, UK

ABSTRACT: The UK aggregates industry faces growing opposition from a wide spectrum of society who are dissatisfied with the ever increasing demands for extraction sites; sites which are often situated in heavily populated or very attractive parts of the country. In addressing these concerns, the Department of the Environment commissioned research into two potential alternative sources of aggregates. Firstly, the importation to the areas of greatest demand of aggregates from remote coastal superquarries in Norway, Scotland and possibly northern Spain. Secondly, the use of recycled mineral and construction wastes. This paper reports on the research into the scope for utilising the various mineral waste sources as recycled aggregates. The already quite considerable utilisation of some waste materials is reviewed and the technical, geographical and economic constraints on greater use are noted. Finally, the paper considers what policy instruments might be utilised to bring about greater use of recycled aggregates and, importantly, what the environmental gains might be.

1 INTRODUCTION

The aggregates industry is still largely local in its orientation. Pits and quarries mostly distribute their products within a radius of about 30 miles. Increasingly over the last 20 or 30 years, however, some areas of the country have become noticeable deficit regions, 'importing' material from neighbouring regions over much greater distances.

There are many reasons why this imbalance in the supply and demand of particular regions should arise, but an important factor is undoubtedly the environmental constraints which operate on production, with their effects working through the minerals planning system. These constraints tend to be greatest in and around the major conurbations where the demand for aggregates is greatest. The pressures of urban development, which is a major consumer of aggregates, conflict directly with the pressures to conserve the countryside and minimize the environmental disturbance of quarrying activities.

The regional pattern of surplus and deficit results in quite substantial inter-regional flows of materials. The south-east, west midlands and north-west regions, in particular, are major deficit regions which import from elsewhere, while the south-west, east midlands and north Wales are the principal 'exporters'.

The obvious solution to the deficit problem has been to mobilise traditional sources of supply of aggregates and, where it has been economic to do so, to transport them over considerable distances. Traditional sources of supply are, of course, sand and gravel, crushed rock and marine-dredged aggregate; but the impact of long-distance transportation has been felt mainly in the supply of crushed rock where economies of scale of production have led to the emergence of inland superquarries which, coupled with productivity improvements on the rail network, have led to quarry-to-railhead distribution systems.

There appear to be limits, however, to the

extent to which established sources of supply can expand to meet the growing deficits of the regions containing major conurbations. Environmental pressures at the major sources of crushed rock supply, particularly in Somerset and in Leicestershire and Derbyshire, suggest that expansion may be more problematic in future.

These environmental pressures and the growing local deficits in aggregates supply have encouraged the search for alternative sources of materials.

One such course is likely to be the development, in remote areas of Norway, Scotland and perhaps northern Spain, of very large sea linked coastal quarries. With the UK this concept has been pioneered by Foster Yeoman's Glensanda superquarry in western Scotland. Another source is the recycling of mineral and construction wastes to produce secondary aggregates. It is this latter source which is the subject of the present paper.

2 THE ROLE OF RECYCLED OR SECONDARY AGGREGATES

The potential environmental gains to be realised from the use of recycling have been recognised for some time. For example, in 1972, the Government introduced a "dual tendering procedure" designed to give greater opportunities for the use of mineral wastes as road fill where such materials were available. In 1976, the Verney Report, "Aggregates the Way Ahead" sought to encourage greater utilisation of mineral wastes as aggregates by recommending a procedure whereby a full cost benefit analysis would be undertaken of different sources of imported fill including secondary materials. More recently, the Department of the Environment's Mineral Planning Guidance Note 6 of 1989 states: "Increased utilisation of wastes could reduce the demand for primary aggregates with benefits of avoiding the dereliction caused by tipping at the same time as reducing the land for extracting natural aggregates".

Nevertheless, policies which would bring environmental costs directly into the decision-making calculations of the building and construction industries and which would alter the balance of materials use in favour of secondary materials have, to date, been strictly limited in their scope and the intensity of their application.

The 1990 Government White Paper on the Environment, "This Common Inheritance" reported the intention on the part of Government to address some of the environmental problems associated with current waste management practice. The White Paper contains proposals which will raise the price of landfill to levels that more accurately reflect the scarcity of the land which may have an effect on the disposal patterns of mineral wastes such as demolition rubble and may make recycling more viable. It further indicated Government's readiness to consider new policy initiatives to internalise environmental costs and pursue the "polluter pays" principle.

It was against this policy context that in early 1990 the Department of the Environment commissioned research from Arup Economics & Planning into the role of secondary aggregates (Whitbread, Marsay, Tunnell, 1991). The main objectives were to:

1. Assess current stockpiles, arisings and utilisation of mineral wastes;
2. Identify technical and economic constraints on greater use of mineral wastes;
3. Recommend policy measures to increase the take-up of secondary aggregates;
4. Establish a framework for environmental evaluation of policy.

The study found that of a total utilisation of aggregates of about 332 million tonnes per annum at the end of the 1980s decade in Great Britain, about 32 million tonnes per annum are derived from secondary materials. This is considerably greater than had previously been assumed. Although there are very many mineral wastes, seven which occur in Britain are by far the most important, either by virtue of the scale of the stockpiles or flows from industrial and construction processes, or of the potential scope for their utilisation.

3 PRINCIPAL ALTERNATIVE MATERIALS

3.1 *Colliery spoil*

This is waste material from the process of extracting deep mined coal, the largest volumes of which are found on tips in Yorkshire, Nottinghamshire and South Wales. There are about 2000 million tonnes of spoil on unreclaimed tips and about 30 million tonnes per annum is still generated for land disposal. The main end use for colliery spoil has been as fill material for road embankments and building sites. The main constraint on its use is its variable quality reflected in a low price, which limits the geographical markets that can be reached.

3.2 *China clay waste*

China clay waste results from the production of china clay in Cornwall and Devon. The industry stockpile is estimated at over 600 million tonnes with about 27 million tonnes of waste arising each year of which around 12 million tonnes is china clay sand. A large proportion of the total waste stockpile is found in the St Austell area, making this the most concentrated mineral waste disposal area in the country. The sand is the only fraction of the waste for which significant secondary uses have been found and about 1.5 million tonnes per annum are used in the local construction industry as aggregate for concrete blockmaking, building mortar, structural concrete, unbound road sub-base and as a general fill material. Increased use of china clay sand could only come about by widening the geographical market area and the study found that this would require considerable financial support to overcome the transport cost barriers involved in accessing major markets.

3.3 *Slate waste*

Between 400-500 million tonnes of slate waste exist on tips, the largest proportion of which is in North Wales but with other sources in Scotland, the Lake District and Cornwall. Around 6 million tonnes are still being generated each year, mainly in North Wales. Until recently there was very limited use of this waste but it has been used as a road sub-base and up to 0.5 million tonnes per annum have been supplied. As with china clay waste the main barrier to further utilisation is the transportation costs necessary to get the material to areas of substantial demand such as the South East and some kind of financial support would be required to overcome these cost barriers, though not nearly as much as in the case of china clay sand.

3.4 *Power station ashes*

In recent years some 75 million tonnes of coal has been burned in coal-fired power stations in Britain each year, which has resulted in the annual production of over 13 million tonnes of ash. Because of the cheaper coal from newer, more productive South Yorkshire and Notts/Derby mines, generation has tended to become focused on power stations in these areas and recent growth in ash disposal has been mainly in South Yorkshire disposal sites.

Virtually all the furnace bottom ash is sold as an aggregate in blockmaking and about 45% of the pulverised fuel ash (pta) which is the main residue. Some 4 million tonnes of pta per annum is sold to the construction industry primarily as a blockmaking aggregate. Other uses include fill and as a cement substitute. Lightweight aggregates are also produced which achieve a very high sales price relative to its general aggregate usage. A major constraint on greater sales of pfa is the location of so many large coal-fired power stations at some distance from the major construction industry demand.

Some of the waste material is used to reclaim derelict sites for which there is an environmental gain, so care would be required if encouragement were given to greater utilisation of this material for aggregates use.

3.5 *Blastfurnace and steel slags*

All blastfurnace slag arisings, some 4 million

tonnes per annum, are fully utilised and some stockpiles of slag from earlier iron manufacturing have been or are being utilised. Only Teesside has unused slag on a significant scale. Of the 2 million tonnes of steel slag produced each year, about 80% comes from a basic oxygen process and the balance from the electric arc furnace process. A constraint on its use is that steel slag contains free lime and magnesia which can cause expansion on contact with water and at present only 10% of BOS slag finds a secondary use, although over 50% of EAF slags are used, mainly for asphalting and other roadstone. The main constraints on further use are the costs of processing the materials which at present are marginally uneconomic. In the Netherlands BOS type slags are also used extensively in road construction.

3.6 *Demolition and construction wastes*

There are about 24 million tonnes of demolition and construction waste arisings per annum in Great Britain and the method of disposal depends very largely on the costs to the demolition contractor. The material is often hauled to landfill sites but the requirement to produce a level building site means that a large proportion of masonry and concrete arisings can be used as fill and in excess of 11 million tonnes p.a. of arisings are re-used.

A number of plants exist which further process demolition and construction wastes into graded aggregates. There appears to be scope for greater use and recycling of these wastes even without policy interventions as landfill costs rise more rapidly in future as a result of measures being implemented under the 1990 Environmental Protection Act. Nevertheless, further encouragement could be given to the establishment of recycling plants in urban areas as this is where the majority of these wastes arise.

3.7 *Asphalt road planings*

A large proportion of all road planings, perhaps as much as 80%, finds its way into some form of secondary use but virtually none is currently recycled back into asphalt for use on road surfaces. In many other countries asphalt road planings are regularly recycled with as much as 50% recycled in Germany, 35% in the Netherlands and 25% in the United States. In these countries central government and most local highway authorities have specifications permitting varying proportions of recycled asphalt to be used in the new mixes. The general conclusion from overseas experiences is that recycling can be economically viable but that the specification is required; the Department of Transport is moving in this direction. Financial incentives have been used in the Netherlands and legal penalties have been used in Germany to encourage the use of recycled materials.

4 ECONOMIC AND ENVIRONMENTAL POLICY APPRAISAL

For most secondary materials however the principal disadvantage associated with further use are the transport costs from source to the principal areas of market deficiency such as the South East. Whilst all aggregates are relatively sensitive to transport costs, there is usually less incentive for mineral waste producers to seek outlets for their material as a secondary aggregate because of the generally lower quality and lower initial prices they can command.

There are other difficulties to be overcome, including customer acceptance of the product, and the need to maintain necessary standards of safety and performance, (where specifications may help) and the study identified a number of actions that government could take to assist with greater utilisation of these materials. Nevertheless, the principal disadvantage faced is economic, arising in the main from transportation costs.

Many of the immediate environmental impacts resulting from the extraction of primary aggregates apply equally to the utilisation of secondary materials. This is because the processes are very similar, namely extraction from an appropriate source, processing often involving noise, dust and visual intrusion, and transportation to market, often using lorries. However, there are longer term environmental gains which favour the utilisation of waste sources for aggregate. These are:

1. The avoidance of the permanent loss of land-related amenity which occurs when aggregates are extracted, for which satisfactory restoration and aftercare can never fully compensate. Similarly, when the mineral wastes are tipped there is also a loss either of land-related amenity or of landfill space which could be saved if more recycling took place;
2. The beneficial use of an otherwise wasted material.

If Government wishes to intervene in the market in some way to bring about greater use of secondary materials, the principal rationale would be to secure these environmental savings.

The study concluded that a package of policy measures would be required to secure maximum use of the secondary materials, but potentially the most effective and direct measures that could be applied are those that will alter the relative prices of primary and secondary aggregates in favour of secondaries.

One method of altering relative prices is a tax or industry-administered levy on primary aggregates and in the following analysis the implications of different levels of tax/levy are followed through. The objective is to provide a means of addressing the question: how much would the environmental benefits be worth.

On the basis of total aggregate demand in 1989/90 of about 332 million tonnes p.a., it was estimated that relative price changes resulting from different levels of tax or levy would bring forth the supply response shown in Table 1.

This assumes that the substitution of secondary for primary aggregates would be the only consequence of the tax/levy. An increase in price might perhaps dampen overall aggregates demand, but this effect is excluded and, for the present purposes of showing the implications of substitution, would not significantly effect the conclusions.

In economic terms, the costs of such a policy are the additional resources required to produce the increased output of secondary aggregates instead of the substituted primary aggregates. So, the economic cost is the cost of the secondary materials less the cost of producing the equivalent amount of primary materials, which is shown in diagram form in Figure 1. The shaded area shows the additional resource cost involved in producing the same total output of aggregates with a tax or levy in place.

Table 1. Effect of price changes on secondary aggregate demand

Price change %	0	15	30	50
Total demand (mtpa)	332	332	332	332
Total sec aggs demand (mtpa)	32	42	55	80
Marginal sec aggs demand (mtpa)	-	10	23	48

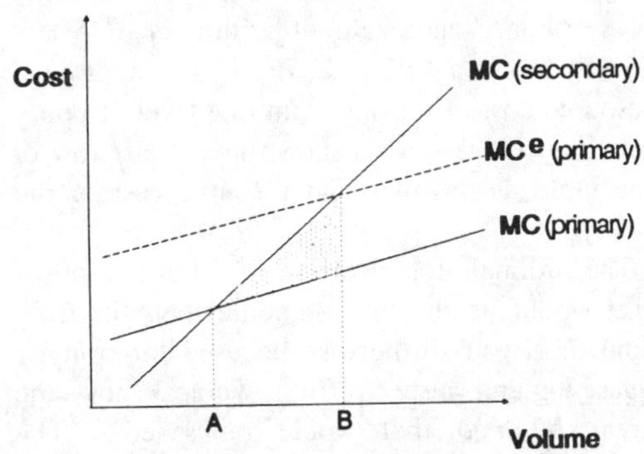

Note: MC = Marginal cost of supply, direct costs only
MC^e= Marginal cost of supply, including environmental costs

Figure 1. Resource cost of applying an environmental tax or levy to primary aggregates

Assuming an average production cost for aggregates of £5.00/tonne, the economic cost of the measure would be calculated as shown in Table 2.

Table 2. The economic cost of a tax or levy on primary aggregates

Price change	15	30	50
Additional sec aggs (mtpa)	10	23	48
Marginal addition (mtpa)	10	13	25
Additional resource cost (£m)	3.75	17.25	60.00
Marginal resource cost (£m)	3.75	13.50	42.75

The additional economic or resource cost given by the shaded area on the figure, is calculated approximately as half the product of the tax rate, the price of primaries and the additional volume of secondary aggregate utilisation eg. (15% x £5.00 x 10 mt p.a.)/2 = £3.75 m. The additional resource cost of moving from one level of policy intensity to the next is shown in the last row of the table as the marginal resource cost of the policy.

The rationale for incurring the above resource cost would be the environmental benefits from land that would otherwise be used for primary quarrying and waste tipping. Table 3 shows the areas of land that would be saved. The calculation assumes that 10 ha of land is saved for every million tonnes of primary aggregates substituted by secondaries and 1.25 ha for every million tonnes of waste disposal avoided.

From these estimates of land area saved, and the resource costs of the policy intervention, it is

Table 3. Annual land saving arising from greater use of secondary aggregates

Price change %	15	30	50
Quarrying land saved (ha)	100	230	480
Tipping land saved (ha)	13	30	60
Total land saved (ha)	113	260	540
Marginal land saved (ha)	113	147	280

possible to assess the cost per hectare which is the (minimum) valuation of the land that would be necessary to justify the policy. This is shown in Table 4.

The finding here is that the higher the tax or levy imposed on the industry producing the primary aggregates, the higher the implied valuation of each hectare of land that is saved. The implied environmental values of the land saved are between £30,000 to £150,000 per hectare when expressed in marginal cost terms. These values (less any allowance for continuity of existing uses such as agriculture) could be viewed as the price of preserving environmental amenity associated with non-quarrying uses of the land.

If a tax or levy on primary aggregates production were introduced as is already the case in Denmark, it would probably be regarded publicly as an environmental tax even if it were not explicitly linked to the imputed environmental gain.

Such a tax or levy would probably achieve greater overall impact and be more compelling if at the same time, grants were made available for restoration and/or for supporting investment in recycling processes.

A scheme administered by the industry itself could probably achieve these ends effectively if,

Table 4. Imputed value per hectare of land saving arising from greater use of secondary aggregates

Price change %	15	30	50
Marginal land area (ha)	113	147	280
Marginal resource cost of policy (£m)	3.75	13.50	42.75
Imputed marginal value/ha (£,000)	30	90	150

for example, the revenues went into an independently administered environmental trust fund. Such a fund could be managed to secure a range of environmentally beneficial objectives. It should not however be used simply to "recycle" funds back into the primary extraction industry as this would defeat the objective of the levy.

A penalty on waste disposal of similar intensity to a penalty on primary aggregate extraction could theoretically have a similar impact on the relative costs of producing secondary aggregates. It could also have undesirable effects such as illegal forms of tipping and could reduce the output of the primary activities such as steel manufacture or china clay production.

5 SUMMARY AND CONCLUSION

To summarise; the recommendation of the study was for Government to give consideration to ways of changing the relative prices of primary and secondary materials in favour of secondaries. It was proposed that this be brought about through a tax or industry levy on the production of primary aggregates. This measure would command greater favour if, at the same time, ways were found to increase expenditure on environmental schemes.

The objectives of a tax or levy would be most likely to be achieved in the context of a package of administrative and regulatory measures including:

1. Greater information on the availability and applications of secondary material use.
2. Planning directives identifying waste material tips that are available for re-use.
3. Support for research into high value applications, including lightweight aggregates.
4. Planning regulations to facilitate the imposition, retrospectively, of land restoration conditions in cases where extraction and waste disposal currently proceeds without any conditions.
5. Support for technical research into the properties of certain materials, such as steel slags.
6. Further development of specifications for the use of recycled aggregates and in particular, demolition and construction wastes for which there is great potential for greater use.
7. Directives to local planning authorities in urban areas to identify sites for recycling plants, particularly for demolition and construction wastes.
8. Grants for modifications to plant and equipment, such as asphalt plants to enable them to accept secondary materials.
9. Grants to overcome transportation difficulties associated with secondary materials in peripheral regions.

Finally, two specific subjects for further research were proposed. Firstly, a more detailed study of the demolition and construction waste industry. This industry has the potential to supply large volumes of graded aggregates in addition to the major contribution it currently makes to general building fill materials. Secondly, the issue of assigning monetary values to environmental amenity.

REFERENCES

Whitbread, M.J., Marsay, A and Tunnell C. 1991. Occurrence and utilisation of mineral and construction wastes, London, HMSO.

Why Recycle?, Rainbow (ed.) © 1994 Balkema, Rotterdam, ISBN 90 5410 367 1

A critical appraisal of recycling the UK roads

A.D.Gill
Colas Limited, Crawley, UK
A.R.Woodside & W.D.H.Woodward
University of Ulster, UK

Abstract

This paper examines the various methods which are presently available in the United Kingdom to enable the recycling of bituminous road materials. It outlines the merits and disadvantages of each system ranging from 'Hot mix off-site recycling at a central plant' to the 'Deep cold in-situ recycling' process. The processes are defined and described and the potential of each operation is assessed by the authors. The paper suggests that in the future a very definite bias will be given towards the use of secondary materials in the sub-bases and road-bases and highlights how a substantial saving of primary aggregates may be made if their place can be taken by secondary materials. The enhancement of secondary materials is considered and the paper concludes by suggesting possible future developments in the light of impending European Standards.

1 INTRODUCTION

Never has the highway industry in the United Kingdom been faced with a greater range of conflicting pressures. Government predictions highlight the need for more roads while the growing "environmental awareness lobby" is calling for restrictions on quarrying and questioning the need for the proposed investment in highway construction.

At present, approximately 320 million tonnes of aggregate are used in the U.K. road building industry every year. This is expected to increase to between 370 million tonnes and 440 million tonnes by the year 2011.

More importantly to meet this demand, imports of aggregate are expected to soar from their present level of 4 million tonnes per year to over 100 million tonnes by the year 2011.

This has created a problem as to where the necessary supplies of aggregate are to come from. To answer this question it may be possible to offer recycling as an acceptable alternative for the future of highway construction in the United Kingdom.

This paper considers the potential for using recycling as a viable option in highway construction. The basic construction and material requirements are outlined first. Sources of materials are then discussed, followed by the factors which need to be considered if recycling is to see future growth in the United Kingdom.

The various methods which are presently available are reviewed, ranging from the traditional view of recycling as shallow in-situ cold-mix and off site/hot-mix, to the more recent environmentally acceptable alternatives such as in-situ/hot-mix and deep-cold/in-situ methods. Finally, the future of recycling is considered in view of the comments proposed.

2 THE LAYERED STRUCTURE OF A ROAD

The structure of a road is made up of a number of layers. Aggregates are required at all levels but both the quality and the cost of the materials used generally increases from the bottom towards the top. This means that specification requirements for the wearing course or topmost layer, are considerably greater than for the bottom capping and sub-base layers.

By building in layers, a very wide range of constructional materials can be used. From a practical point this relates to reductions in cost if abundant local low quality materials can be used to provide the large amount required for the lower layers. Although the quantity of materials is less with higher layers their total cost may be greater as they may have to be transported considerable distances should suitable local supplies not exist.

For each layer, different specification requirements are needed as the different layers perform different functions. This ranges from the ability to withstand the polishing and attrition caused by trafficking to the distribution of stresses in the lower layers, i.e. it is a case of "horses for courses".

In this context, it would be very attractive should secondary and other types of recycled materials be shown to perform to the same standard as traditional sources now in use.

3 SOURCES OF CONSTRUCTION MATERIAL IN THE UNITED KINGDOM

The term construction material has been used to include the wide range of materials now seen to be available for road construction. They include:-

3.1 Primary or virgin sources of aggregate produced specifically for road construction

These are the traditional sources of aggregate, and have been selected over time as being suitable to give the necessary levels of in service performance. However, they have a problem in their distribution which is dependent on the geology of the United Kingdom. This means that there are few sources of the required "hard rock" to be found in South-east England where demand is greatest. Rather, aggregates that have been used tend to be found further to the North and West, for example, in the Peak District and North Wales. Indeed, a number of Northern Ireland quarries now find it profitable to export aggregate, both high quality surfacing chippings and lower quality sub-base material, into South-east England.

3.2 Low quality aggregates

These are sources which the highways industry may have tried in the past and have been shown not to possess the necessary levels of performance when used in road construction. As a result, such materials may now only be used as filler or as sub-base.

3.3 Secondary materials

These have been created as a by- product of another industry. Examples include slate waste from North Wales, colliery waste from the coal mining areas, china clay sand from Cornwall, fly ash created by electricity production and slag from the steel industry.

3.4 Processed demolition waste

This is an industry of growing importance, especially in parts of Europe where natural resources may be even more scarce than in the U.K. Many quarries are now recycling materials such as concrete rubble from building demolition.

This practise is being enforced with levies imposed against the use of quarried virgin aggregate. In its favour, the Dutch have for example found that when used as sub-base material, a recementation process occurs which serves to increase the strength of the layer.

Current research being carried out at the University of Ulster in conjunction with Building Research Establishment is assessing the potential of such materials.

3.5 Surface planings and bituminous/ unbound aggregate dug-out from pre-existing highway construction

This offers the greatest potential for recycling as the raw material is already on-site and only requires a suitable type of recycling process to produce a material which will perform in use. As a result, the existing road is being used as a horizontal quarry.

4 FACTORS TO CONSIDER ABOUT RECYCLING

Although it is possible to say that there are many hundred's of millions of tonnes of potentially recyclable material available, the fact of its existence does not automatically warrant their use in a highway's construction. Other factors need to be taken into account before what is perceived as the environmentally acceptable alternative of recycling is adopted by the industry. Some of these factors will now be considered.

4.1 Location

Whilst one can quote mountains of slate waste in North Wales, such material still has to be transported considerable distances to where demand for bulk aggregate is greatest, i.e. predominantly the South-east of England. In practise this implies that greatest use of this type of material is restricted to a localised market.

In contrast, there are many miles of the British road network now undergoing advanced stages of structural or surfacing failure. Recent figures suggest that the number of roads with less than one years life expectancy may be over 15% of the total road network. It is these failed roads which could provide abundant on-site sources of material which favour in-situ recycling processes.

4.2 Transportation costs

These are a major factor to consider in the future use of low-grade and secondary materials. Other than for localised use, transportation over distance only becomes attractive once the value of the product is raised, such as in the case of high skid-resistant surfacing aggregate.

4.3 Traditional experience

The use of non-traditional sources of aggregate, such as those proposed for recycling, do not have the long term track record of traditionally used sources of aggregate.

This aspect of going into the unknown, is one which the British roads industry has traditionally been loath to risk.

4.4 Long term performance

Whether due to pressure from Europe, environmentalists, or growing concern about rapidly depleting sources of aggregate, governmental bodies such as the Department of Transport and the Department of the Environment are now investing considerably into research about the use of these materials so that the necessary in service experience is obtained.

4.5 Material durability

If aggregate is to be used in the context of the modern road, then it must be expected to possess adequate long term performance properties. There is no point in using recycled materials in a situation where in service conditions cause premature structural failure. Consider the example of recycling a bituminous material which had prematurely failed due to aggregate durability or soundness reasons. Unless detected by adequate screening and testing procedures, the resulting structure is simply premature failure waiting to occur.

4.6 Modern environmental pressures

This is a veritable minefield with the quarrying and highways industries placed in the middle of conflict between those who forecast the need for increased growth and those who would impose greater restrictions on all future growth. Hard rock aggregate has traditionally been quarried in what are now public areas, such as the Peak District National Park and so future development is under much debate. An alternative has been to go away from the public and develop the remote super-quarry such as that at Glensanda off the West coast of Scotland where the in-side of a mountain is quite literally being quarried away. Having a coastal site, handling and transportation costs may be minimised.

5 USES OF RECYCLED MATERIALS IN HIGHWAY CONSTRUCTION

There are two main uses for recycled materials:-

5.1 Capping and sub-base materials

As the specification requirements for these materials are quite low, there is a great potential market for such recycled materials, either on their own or in combination with primary materials such as virgin crushed rock aggregate.

5.2 Roadbase and surfacing materials

As the specification requirements for these layers are much higher, this necessitates the raw recycled material to possess a higher level of performance. Candidates for this type of use are surfacing planings which should contain a relatively high quality aggregate. However, as mentioned previously, unsound aggregate is a problem that must be considered.

6 ROADBASE AND SURFACING RECYCLING PROCESSES

The use of recycling in these layers usually, but not always, requires the material to be bound with bitumen or cement. It is possible to categorise the different types of recycling process as used for roadbase and surfacing layers as follows:-

Hot-mix/off-site

Cold-mix/off-site

Shallow hot-mix/in-situ

Shallow cold-mix/in-situ

Deep cold-mix/in-situ.

Generally, in terms of cost, the in-situ processes are to be favoured as they do not require extra transport, handling and processing - if, it can be shown that the in-service performance is comparable to the traditional hot-mix/off-site process.

Each of these process types will now be defined and discussed. The potential of each will also be given.

6.1 Hot-mix/off-site processes

This is the traditional type of recycling of which there is still a limited knowledge in the United Kingdom. In this process, existing materials are removed by planing, transported to a hot-mix plant and then reprocessed with virgin aggregate and bitumen to comply with specification requirements for hot-mix materials such as Hot Rolled Asphalt.

In Northern Ireland, this method has been used by the Department of Environment Roads Service, for a number of motorway and dual-carriageway re-surfacing contracts. The first was in 1988 on the M1 motorway and involved the use of Hot Rolled Asphalt planings applied to virgin aggregate and bitumen. It was found that the recycling mixing process required careful control to provide a satisfactory end-product. Initial trials proved that mixes containing up to 50% recycled material could be used successfully.

However, the future of this type of recycling in Northern Ireland is limited due to the abundance of cheap, high quality virgin aggregates. But, with the use of a modern, mobile hot-mix plant on-site, this method may prove to be a viable option in mainland Britain where surface aggregate costs are significant.

As regards its future, the Department of Transport currently only permit the use of mixes with up to 10% although levels of 50% are used elsewhere in Europe. In the long-term, European environmental and health concerns may not allow this type of recycling process to reach its full potential, but may however favour the growth of cold-mix processes.

6.2 Cold-mix/off-site processes

This is similar to hot-mix/off-site in that it involves recycling at a central plant. The exception is that the process involves the use of cold mixing with either one or a combination of foamed bitumen, bitumen emulsion, cement and lime.

As regards the only operational plant in the United Kingdom, this is now operated by R J Maxwells and has to date concentrated on the use of foamed bitumen stabilisation. This, when mixed with bituminous planings produces a material suitable for roadbase use. As such this process has significant potential within the U. K marketplace. However, it has also been trialed as a permanent re-instatement material under the U.K's HAUC regulations which may further extend its market potential.

6.3 Shallow hot-mix/in-situ processes

This may be termed a surface re-generation process for existing wearing course materials such as Hot Rolled Asphalt.

The process first involves heating the roads surface layer, scarifying followed by reshaping and then reinforced by a thin overlay of new asphalt. Typically this is 20 - 25mm thick and is heat welded to the old material. The total depth of treatment is about 50mm with cost savings of 15 - 20% claimed. Examples of this type of process include that known as Repave.

Due to the size of plant involved, this process has tended to be restricted to major roads. But, as trunk roads and motorways account for only about 4% of the United Kingdom's total road network, the processes potential expansion must be restricted. However the process is beginning to show favour in Eastern Europe.

6.4 Shallow cold-mix/in-situ processes

This process, commonly known as Retread, has been in service for over 50 years in this country. It was originally introduced as a relatively cheap method of repairing badly damaged roads during the war.

The fact that it has survived as long, is testimony to its value within the road maintenance industry in this country.

Retread involves firstly the scarifying and reshaping of an existing road or footway surface. Once completed, virgin aggregate may be added to reprofile the road surface, alternatively excess aggregate may be removed.

After the desired profile has been achieved, bitumen emulsion is applied using a spray tanker. This is harrowed in to the full 75mm depth of the retread layer to ensure an even mix. This is then followed by compaction of the layer.

Finally a surface dressing is applied using between 3mm - 14mm chippings to give adequate texture depth to the surface.

Depending upon the type of emulsion used in the final dressing, and if the site is to be subsequently overlaid, a further surface dressing may be required in 9-12 months to finally seal the surface.

This method of in-situ cold recycling is appropriate for the rejuvenation or reshaping of residential and generally lightly trafficked roads. The Retread process has been shown to be a cost effective alternative to planing out and adding a new overlay. Giving a claimed cost saving of between 25 - 35%.

As it uses a "cold emulsion" it also has the advantage of being attractive from both a Health and Safety viewpoint as well as to the environment.

6.5 In depth cold-mix/in-situ processes

As the name implies, this process treats the road to a far greater depth than does the shallow cold-mix/in-situ process. This type of process can recycle an existing road surface to a depth ranging from 150mm to 300mm.

This process involves pulverising the existing road surface to a depth of up to 300mm. This material is then compacted and reshaped.

Excess material is removed at this stage. Once the desired profile has been achieved the material is rotovated again, during which time bitumen emulsion, foamed bitumen and/or cement will be added in pre-determined quantities and thoroughly mixed throughout the layer.

Again the layer is compacted and shaped before being sealed with sprayed bitumen emulsion and sealing grit. Typically this layer is then overlain with some other material to provide a new running surface.

As well as an enhanced speed of operation the in-depth recycling process typically offers a cost saving of up to 40% and an energy saving of up to 10% when compared to traditional methods, since the existing road is being used as a horizontal quarry.

Its principal advantage is that is very flexible depending upon what is being recycled. Due to its significant financial and environmental benefits, and the pressure now being placed upon local authorities to recycle, this type of process must have potential in the future. Recent research at the University of Ulster has shown this material was capable of out performing "virgin" material.

7 CONCLUSIONS

To be critical about the future of recycling in the United Kingdom, it is possible to conclude:-

Modern roads require materials that can provide the necessary levels of long term performance.

The high demands for surfacing aggregate are generally too great for secondary recycled materials.

Both the Department of Transport and the Department of the Environment desire to see increased growth in the use of recycled materials for structural and/or foundation layers of existing and proposed roads due to the quantity of materials required.

Hot-mix recycling is already acknowledged by the Department of Transport as a viable option, but has problems such as extra cost and quality control.

In the debate of hot versus cold processes, cold mixes are more attractive due to their environmental acceptability.

There would appear to be good, long term prospects for the in-situ cold recycling processes.

Off-site processes still involve considerably more energy expending activities - such as road excavation, planing and importation of the recycled materials.

The authors wish to acknowledge the help and co-operation received from their colleagues in the preparation of this paper and would emphasise that the views expressed are entirely their own and are not necessarily those of the University of Ulster or Colas Limited.

Why Recycle?, Rainbow (ed.) © 1994 Balkema, Rotterdam, ISBN 90 5410 367 1

Recycling of colliery minestones into useful engineering materials

A.Chrysanthou
Department of Materials Engineering and Materials Design, University of Nottingham, UK

ABSTRACT: An investigation was carried out to examine the potential for producing useful engineering materials from colliery minestones. Mullite-SiC composite powders were formed at temperatures below 1573K while Al_2O_3-SiC was obtained at higher temperatures. Sialon ceramics were also prepared by carbothermic reduction in $N_2/5\%H_2$ mixtures. The nature of the raw materials had a marked influence on the kinetics, the product morphology as well as the reaction mechanism. A feasibility study showed that the preparation of porous ceramics was also possible.

1 INTRODUCTION

Colliery minestones are minerals which are associated with coal and are a waste product of coalmining. Large quantities of minestones are produced worldwide and have to be deposited at landfill sites or discarded into the sea. Pressure from environmental groups may result in governments imposing a special dumping tax on companies using landfill sites. The issue of recycling minerals including colliery minestones is therefore becoming a priority from both the environmental and economic point of view. The study of the production of advanced materials from low cost minerals or waste products has been of interest since the late 1970's. Lee and Cutler (1975) showed that it is possible to prepare silicon carbide (SiC) from rice hulls since they contain silicon dioxide (SiO_2) and cellulose which acts as a source of carbon. Subsequent studies(Milewski 1985, Patel 1989, Sharma 1984) have resulted in a well-established process as well as a patent (Patel 1985). An investigation by Patel and Kumari (1990) who used sugarcane leaf and rice straw as raw materials led to the formation of SiC particles and whiskers. Sialon ceramics were also obtained by carbothermically reducing cheap raw materials like kaolinite and mullite under a nitrogen atmosphere (Bergaya 1992,Bigay 1988, Le Doussal 1988,Riley 1988). The impurities present were reported to have some effect on the mechanical strength of densified sialon powders (Le Doussal 1988).

The object of this study was to examine the feasibility of using colliery minestones as raw materials for the preparation of advanced ceramics including Al_2O_3-SiC composites and sialons. Al_2O_3-SiC composites are widely used for cutting tool applications while sialons are important refractory materials used in the ferrous industry.

In recent years there has been an increasing interest in porous ceramics which are now widely used as catalyst supports and filters (Chan 1991, Gonzalez-Oliver 1988). Porous ceramics are produced by the addition to ceramic powders of organic materials which are burnt to leave behind a structure of pores. It was decided to investigate the potential of making porous ceramic materials from minestone M by burning the carbon contained in the minestone.

2 EXPERIMENTAL

Al_2O_3-SiC composite powders were produced in a horizontal tube furnace under a high purity flowing argon atmosphere. Flowing nitrogen gas containing 5% hydrogen was used for the preparation of sialon powders. The extent of reaction was monitored by measuring the weight change of samples at specific intervals. All samples were reacted in the form of loose powders. Raw material and product characterisation was carried out by means of

Differential Scanning Calorimetry (DSC) X-ray diffraction (XRD) and scanning electron microscopy (SEM).

3 RAW MATERIALS

Two kinds of minestones which were supplied by the British Coal Corporation were chosen for this study. Both were clay minestones and belonged to the kaolin group. It was verified by XRD that kaolinite, $Al_2Si_2O_5(OH)_4$ was the major phase present. The minestones also contained amounts of SiO_2 in the form of quartz and minor constituents including K_2O, MgO, Fe_2O_3 and TiO_2. The chemical compositions of the minestones labelled M and N are listed in Table 1.

Table 1: Minestone chemical composition (weight%)

	Minestone M	Minestone N
SiO_2	28.59	51.35
Al_2O_3	19.88	29.85
Fe_2O_3	1.97	2.84
K_2O	2.48	2.31
TiO_2	0.39	0.98
MgO	0.76	0.89
C	25.00	-
Moisture content	20.40	12.55

A comparison of XRD peaks showed that minestone M contained a greater ratio of $Al_2Si_2O_5(OH)_4$ to SiO_2 than minestone N. In Table 2 are presented results of the percentage weight loss on ignition in air at various temperatures as obtained by heat-treatment in a muffle furnace.

Table 2: % weight loss on ignition

Temperature K	473	973	1373
Minestone M	5.31	45.38	45.35
Minestone N	2.84	12.06	12.55

Minestone M contained 25% carbon in the amorphous state which acted as the reducing agent for the production of Al_2O_3-SiC composites and sialons. Fig. 1 is a typical micrograph of minestone M showing narrow seams of carbon embedded in a kaolinite particle. Minestone N contained no carbon so carbon black was added to it as a reducing agent. For the purpose of producing porous ceramics only minestone M was used.

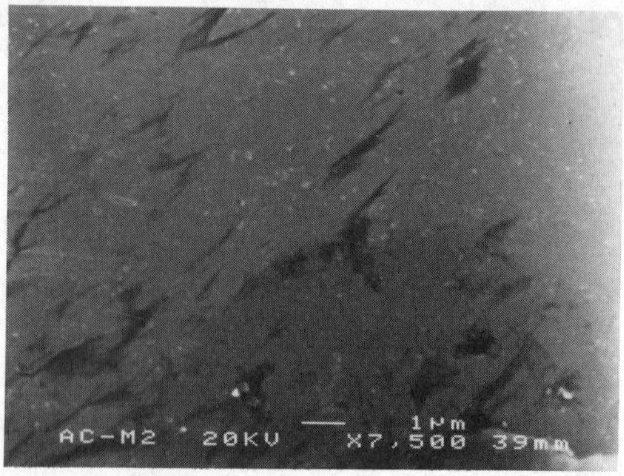

Fig.1 Micrograph of minestone M

A combination of DSC and XRD studies indicated that several changes took place prior to the reduction of both minestones. At 808K $Al_2Si_2O_5(OH)_4$ decomposed exothermically to metakaolin ($Al_2O_3.2SiO_2$) and water. Metakaolin then decomposed to mullite ($3Al_2O_3.2SiO_2$) and SiO_2 at 1223K. These observations were in contrast to the work of Brindley and Nakahira (1959) who have claimed a different reaction sequence leading to mullite at 1673K.

4 RESULTS AND DISCUSSION

4.1 Preparation of Al_2O_3 - SiC powders from minestone M

The minimum temperature for the carbothermic reduction of this minestone was observed to be 1523K. Only SiO_2 could be reduced at this temperature while a temperature in excess of 1573K was required for the reduction of mullite. The available thermodynamic data (Turkdogan 1980) predict that the overall reaction

$$SiO_{2(s)} + 3C \text{ graphite} \rightarrow SiC_{(s)} + 2\,CO_{(g)}$$

takes place above 1781K. The reaction mechanism is in two steps:

$$SiO_{2(s)} + C \rightarrow SiO_{(g)} + CO_{(g)} \text{ step 1}$$

$SiO_{(g)} + 2C \rightarrow SiC_{(s)} + CO_{(g)}$ step 2

According to the thermodynamic data step 1 will not take place below 2000K. The much lower temperatures of reduction observed in this work are due to the high reducing ability of the coal contained in the minestone as compared to graphite. This implies that much higher partial pressures of gaseous SiO occur thus making both steps of the reaction mechanism possible at lower temperatures than expected. The same is true for the reaction

$$(3Al_2O_3.2SiO_2)_{(s)}+3C\rightarrow 1\tfrac{1}{2}Al_2O_{(s)}+SiC_{(s)}+2CO_{(g)}$$

which is predicted to occur above 1817K. The final products above 1573K consisted of β - SiC and Al_2O_3, while between 1523K and 1573K, β - SiC and mullite were observed. Thus, a choice of products was available by controlling the temperature of reaction. An indication of the time required for complete reaction at various temperatures is shown in Table 3.

Table 3. The effect of temperature on the time for complete reaction.

Temperature K	Time for complete reaction (hours)
1573	16
1673	5
1723	3

Two different types of SiC particles were detected by SEM. SiC which was obtained by the reduction of mullite was associated with Al_2O_3 in the product microstructure. This is evident in fig.2 which shows a particle produced at 1723K the outer part of which is SiC while the core contains Al_2O_3 and unreacted carbon. The reason for this observation is related to the nature of the minestone. Since prior to reaction individual particles contain mullite, SiO_2 and carbon, all reactions occur within the particles without the need of diffusion to or from outside. The result of this is the production of composite grains of Al_2O_3 and SiC. This unusual microstructure is very interesting as it is likely to have a unique set of properties. The presence of carbon within the oxide particles suggests that shorter diffusion distances may be required by the reducing agent

Fig.2 Particle from the reduction of minestone M with SiC at the periphery and an aluminium oxide core containing unreacted carbon.

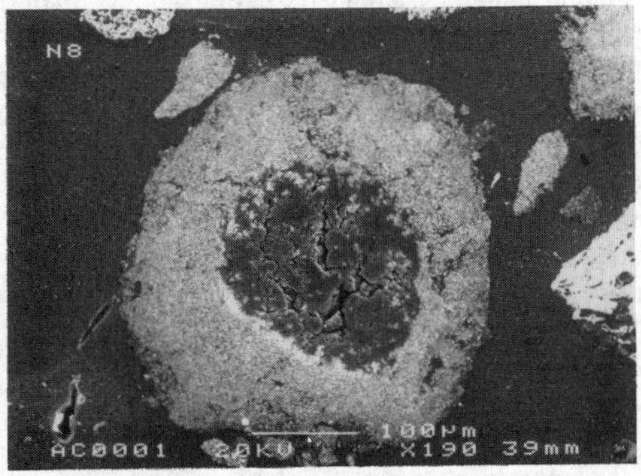

Fig.3 SiC around carbon black obtained by reduction of minestone N at 1723K.

and this may be partly responsible for the faster rate of reaction for minestone M than N.

The other type of particles which was observed contained SiC only. These particles were found less frequently in the microstructure. They were thought to originate from free SiO_2 which was present in the as-received material.

4.2 Reduction of minestone N

Complete reduction of minestone N at 1723K was achieved after about 7 hours indicating that it reacted more slowly than minestone M. The reason for this was thought to be associated with the nature of the reducing agent used (as explained in the previous section).

The product morphology had a marked contrast to that obtained from minestone M. Almost all the SiC particles were totally free from Al_2O_3 and generally formed around a core of carbon black. A typical such microstructure is shown in fig. 3. In this case contact between mullite and carbon black exists on the external surface of grains. As a result gaseous SiO had to diffuse out of them to encounter more carbon black to form SiC. A SiC layer thus formed around carbon black resulting in a morphology where SiC and Al_2O_3 were separate from each other. In all other respects the two minestones behaved the same way.

4.3 Effect of "impurity" oxides

One of the main concerns related to the use of colliery minestones for the production of advanced ceramic materials would be the behaviour of the "impurity" oxides. The alkali oxides present are likely to be reduced at the processing temperatures with evolution of Na and K. On the other hand CaO and MgO are thermodynamically stable at these temperatures and will not be reduced by carbon. Indeed their presence may be beneficial as they are expected to promote the densification of these materials during sintering. Examination of XRD patterns of reacted samples showed that the only significant peaks other than Al_2O_3 and SiC were due to iron which is known to catalyse the carbothermic reduction of SiO_2 (Lee, 1975).

The presence of metallic iron is also likely to increase the toughness of the composite. Observations showed that iron was associated with Ti(O,C) in the microstructure. The latter was produced by the reduction of TiO_2 and is a hard ceramic material.

4.4 Production of sialon from minestone M

Sialon, $Si_3Al_3O_3N_5$, was successfully produced by carbothermic reduction in a $N_2/5\%H_2$ gas mixture at temperatures between 1523K to 1623K. The sialon phase appeared to be obtained directly from mullite. At 1623K it took approximately 8 hours for the completion of the reaction. Fig. 4 shows a micrograph of $Si_3Al_3O_3N_5$ produced at 1623K. Analysis showed that these agglomerates contained about 1.5% MgO which formed glassy silicate phases to promote sintering. The presence of oxides of Mg and Ca was therefore deemed to be beneficial as they aided the consolidation of the sialon. Conditions above 1673K were more reducing and resulted in the formation of AlN and Si_3N_4 as the final products.

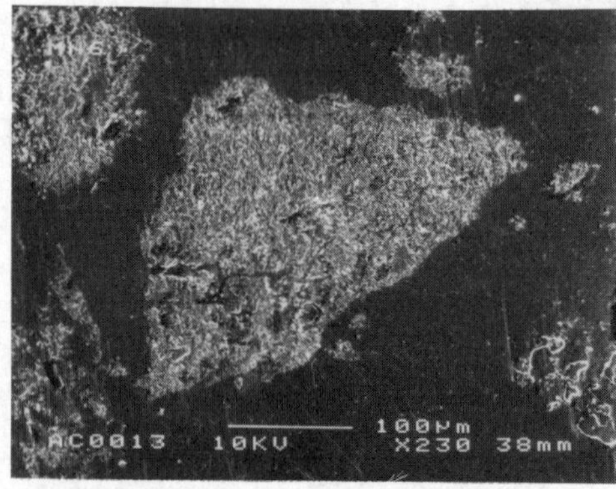

Fig.4 Sialon from minestone M at 1623K.

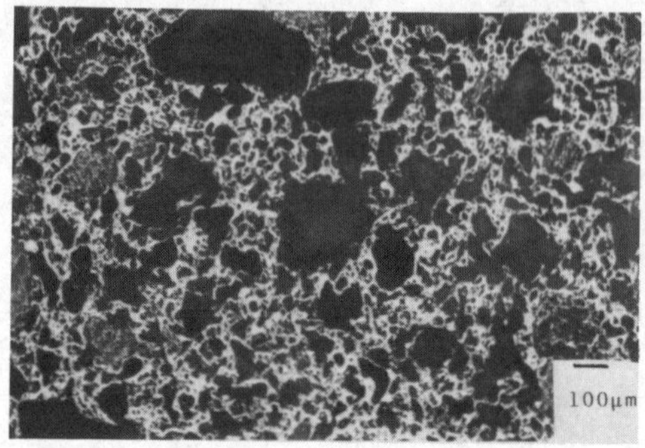

Fig.5 Porous ceramic from minestone M by firing at 1473K.

4.5 Production of sialon from minestone N

$Si_3Al_3O_3N_5$ was also obtained by carbothermic reduction of minestone N in $N_2/5\%H_2$ atmospheres. This was only possible at temperatures between 1573K and 1673K indicating that slightly higher temperatures were required for minestone N than M. XRD analysis showed that $Si_3Al_3O_3N_5$ was formed from mullite via $Si_{12}Al_{18}O_{19}N_8$ and $Si_6Al_6O_9N_8$. At higher temperatures conditions became more reducing such that AlN and Si_3N_4 were the eventual products.

4.6 Porous ceramics

Minestone M was mixed with water and pressed into compacts measuring 45 x 10 x 8mm.

After drying, the compacts were heat-treated at 1123K in air to burn off the carbon and then fired at temperatures between 1473 - 1673K for 5 hours. It took approximately 5 hours to burn off all the carbon at 1123K. The final ceramic contained mainly mullite with a small amount of SiO_2. Some evidence of glassy phases was also found.

Porous ceramics of continuous open porosity were obtained. The internal structure of a porous ceramic produced by firing at 1473K is shown in fig. 5. The pore size distribution was fairly wide varying from 10μm to 600μm. This was thought to be due to the wide particle size distribution of the starting material as well as the non-uniform size of the carbon specks. The use of a finer raw material is expected to give more controlled porosity with finer pores. The porosity remained relatively constant at about 76%. The 3-point bend strength was observed to increase with temperature as shown in Table 4. Na_2O, K_2O, CaO and MgO formed glassy phases with SiO_2 and promoted sintering. The increase in bend strength with firing temperature was due to glass formation becoming easier at higher temperatures.

Table 4. The effect of sintering temperature on the bending strength.

Sintering temperature, K	3-pt.bend strength $Kgfcm^{-2}$
1523	24.3
1573	36.2
1625	44.4
1673	49.1

CONCLUSIONS

The investigation has shown that it is possible to produce ceramic composite powders and sialon from colliery minestones. Between 1523K and 1573K mullite-SiC powders were obtained while above 1573K Al_2O_3 - SiC could be formed. In the presence of nitrogen, sialons were prepared at temperatures between 1523K and 1623K for minestone M and 1573K to 1673K for minestone N. It is also feasible to produce porous ceramics based on mullite by burning off the carbon content of minestone M. A fine minestone particle size would be required in this case to give a finer and more controlled porosity. The presence of MgO and CaO was observed to be beneficial as they promoted the sintering of ceramic particles.

REFERENCES

Bergaya F, F Kooli & J F Alcover 1992. Synthesis of SiMON, ceramics from various clays and different sources of carbon. J Mater Sci 54: 2180-2186.

Bigay Y, B Cales & J P Torre 1988. Sintering behaviour and thermomechanical properties of sialons obtained from low cost raw materials. In H Nosbusch & I V Mitchell (eds), Technical ceramics: 51-58. New York: Elsevier.

Brindley G W & M Nakahira 1959. The kaolinite-mullite reaction series. J Amer. Ceram. Soc. 42: 314-318

Chan K K and A M Browstein 1991. Ceram. Bull 70: 703

Gonzalez-Oliver, C J R., M Schneider, K Nawata and H Kusano 1988. J Non-Cryst. Solids 100: 274

Le Doussal H 1988. Sintering behaviour and thermomechanical properties of sialons obtained from low-cost raw materials. In H Nosbusch & I V Mitchell (eds), Technical ceramics: 59-64. New York: Elsevier

Lee J & I B Butler 1975. Formation of silicon carbide from rice hulls. Ceram. Bull 54: 195 -198

Milewski J V, P D Gac, J J Petrovic & S R Skaggs 1985. Growth of beta-silicon carbide whiskers by the VLS process J Mater Sci. 20: 1160-1166.

Patel M, C B Raju, A K Ray & A Karera 1985. Indian Patent filed INF/91/85 Pat.

Patel M & A Karera 1989. SiC whiskers from rice husk: role of catalysts. J Mater Sci Lett 8: 955-956.

Patel M & A Kumari 1990. Silicon carbide from sugarcane leaf and rice straw. J Mater Sci Lett 9: 375-376

Riley F R & H Mostaghaci 1988. Sintering behaviour and thermomechanical properties of sialons obtained from low-cost materials. In H Nosbusch & I V Mitchell (eds), Technical ceramics: 65-71. New York: Elsevier.

Sharma N K, S Williams & A Zanguwill 1984. Formation and structure of silicon carbide whiskers from rice hulls. J Amer Ceram Soc 67: 715-720.

Turkdogan E T 1980. Physical chemistry of high temperature technology. New York: Academic Press.

Why Recycle?, Rainbow (ed.) © 1994 Balkema, Rotterdam, ISBN 90 5410 367 1

Classification of mineral wastes from mining and preparation of coal

M.Ya.Shpirt
Institute of Mineral Fuels, Moscow, Russia

A.K.M.Rainbow
MES International, Bedlington, UK

OUTPUT AND COMPOSITION AND PROPERTIES OF WASTE FROM COAL MINING AND PREPARATION

Estimates of the development of fuel / power resources of the C.I.S and for the majority of other countries under-line an increase in the utilization of fossil fuels in the next few decades. This will lead to an increase in coal mining in countries with large reserves; in others such as the U.K mining will decline.

Because of mining and geological conditions the amount of waste associated with coal mining will increase. This has certainly been the tendancy over the last 10 years or so.

In 1980 the annual output of waste in the large-scale mining countries exceeded 40-50 million tonnes, for instance Poland produced 47 million tonnes, West Germany and United Kingdom each produced in the order of 60 million tonnes and the United States waste production exceeded 90 million tonnes. Those countries whose coal production has increased have also significantly increased their waste production; however countries like Poland and U.K have reduced their over-all production due to the decline in coal production.

The cost of handling the wastes are increasing enormously - the deposit of wastes in spoil heaps effect the environment in all its three phases, air, land and water.

It is incumbent on mining authorities to increase the utilization of what is often a valuable raw material.

The utilization of mining wastes reduces the cost for the industry and reduces environmental impact.

Deposits of coal mining waste whether situated in close proximity or considerable distances apart consist of a great variety of rock types, shapes and sizes. These in turn have a wide variety of lithological and mineralogical characteristics, properties and contents of organic matter, sulphur content, micro-elements and other characteristics, Therefore mining waste as a whole can be described as having large variations in their physical and chemical properties.

It is important to classify in particular ways if the material is to be of any value for industrial use.

In order to choose appropriate parameters for different end-uses it is necessary to identify where these materials could be used satisfactorily.

THE MAIN AREAS OF RESEARCH AND THE LEVEL OF UTILIZATION OF MINING WASTE IN FORMER SOVIET UNION

The output of coal totalled 1500 million tonnes from opencast coal sites and 100 million tonnes from deep mines. The annual production of waste emanating from coal production and power plants is estimated to be 90 million tonnes.

It is also estimated that about 50% - 45 million tonnes - of the mining waste is used for the reclamation and levelling of land which has suffered severely from mining subsidence. Mining subsidence resulting in a land lowering of over 15 metres has been reported. Coal mining wastes are used for the construction of highways and the sealing of disused mining shafts.

Mining wastes are used for the manufacture

of bricks, incorporated into cement manufacture and processed to recover limestone aggregates.

By employing separators more than 1.5 million tonnes of conditioned fly ash is recovered from power plants. These recovered materials are used in the manufacture of ceramics (mainly wall tiles) lightweight building blocks and in some cases as components of highway embankments.

Current research in Russia is directed on transforming coal wastes from raw materials into production of alumina and porous lightweight aggregates.

It has been possible from results obtained from other countries on the classification of coal wastes to enable parameters to be established for the raw materials to be used in industrial processes.

MAIN PRINCIPLES OF CLASSIFICATION

Some properties of mining waste ie. size and moisture content, are to a large extent dependant on the methods used to produce coal. Therefore it is suggested that the differences be identifyed from the outset ie. identifying the source from which they originate. This is referred to as parameter A.

PARAMETER A

l. Mining waste - opencast or underground
ii. Preparation waste.
iii. Heat treatment waste.
iv. Waste from tips.

Waste from iii. and partly from iv. can be regarded as thermally prepared waste since they will have been altered by associated heating conditions.
One of the significant features in groups i. and ii. is the organic matter content. Research indicated that the resulting waste will have very similar characteristics especially if the organic carbon content exceeds 4%.

These classifications are only qualitative and imprecise for safe clarification therefore an quantitative system needed to be improvised.

It is considered that compared to the other groups; group ii., the waste resulting from the coal preparation plants wast the most stable in its composition.

The material can be characterized using some measurable parameters which can identify its suitability or otherwise for utilization; ie. grain size, chemical and physical composition etc.

Grain size can be sub-divided on the basis of the sizes governed by the flotation process. These could be : large, more than 13mm, medium, 1mm-13mm and fine, less than 1mm. Each sub-division will have similar moisture contents since moisture content is related to grain size. The chemical and physical properties will be largely dependent on the chemical and physical properties of the organic material present. Generally the material can be regarded as falling into three groups - clay - sandstone - carbonates and it is suggested that these groups represent another category for classification purposes. Should a more detailed description be needed the clays may be further sub-divided ie. into predominately kaolins, montmorillonites and so on.
In most cases coal wastes will contain a mixture of different clay minerals but it should be possible to decide to which sub-group they should be allotted.
These parameters are the B parameters.

The existence of organic matter is the most important feature of waste from coal mining in comparison with other raw materials . This index is very important for both the evaluation of coal wastes as a raw material for thermal treatment and for use as a construction material. The parameter that establishes the organic matter is the organic carbon content and this parameter will be known as the C parameters.

The classification system needs to recognise the degree of coalition of the organic matter and this parameter is the D parameter. The D parameter will allow forecasts to be made on the speed of burning of the organic matter, the quantity of resinous materials exuded as well as the design of the heat treatment plant.
It is suggested that such a parameter, like many of the other parameters identified, could be further divided into say, organic content of the fuel mass or the reflectance of the vitrinite.

An equally important constituent in coal mining waste is the sulphur content (S). Therefore S is a parameter. The level of iron compunds (Fe) may influence the potential utilization of the waste and the proportion the paramater Fe, Ferric oxide, in the ash could be a useful parameter in defining acceptable levels of iron compounds.

THE SYSTEM OF CLASSIFICATION AND ITS USE FOR CHARACTERIZATION OF WASTE FROM COAL MINING AND PREPARATION AS A RAW MATERIAL FOR INDUSTRIAL UTILIZATION

Table 1 can be used as a system of classification which can speedily dtermine the suitability or otherwise of coal mining wastes on the basis of laboratory tests.

The proposed system requires six parameters (A,B,C,D,Fe,S) for each type of coal waste to be dtermined before a recommendation can be offered on the possible and technical advisable utilization.

A review of the influences of these parameters show that A - source of coal mining wastes - would probably, in most situations, favour A2 - coal preparation waste. In Russia they have concentrated on developing a reliable method for testing A2 type material. In the UK the preferred material would be A4a and b since this material has already been subjected to an A2 operation and the chemical and physical properties will have greatly stabilized whilst on the tip.

The process of using heat treated wastes is well advanced.

It is advisable to perform all six parameter evaluations in order to establish a particular utilization potential. For example, on lithological composition (B) clay coal wastes (Bl) can be recommended as a raw material for the manufacture of porous filters for lightweight concretes and ceramics. If consideration is given to parameters C,D,Fe and S their potential use could be extended into the production of silicone -aluminium alloys or by considering C and Fe an evaluation of the abrasiveness could be made - for highway and engineering earthworks. Parameters D and S may prove to be the limiting criteria.

From clay wastes, there is potential for the utilization of the wastes of the kaoline group (Bla) because they have a high potential for the manufacture of fine and acid resistant porous filters.

Carbonate coal wastes (B3)can be recommended as a raw material for the manufacture of binders if we are to include in this evaluation of C,D and S parameters they could be used as an aggregate in highway construction.

It is always necessary to measure the carbon content (C) when looking for potential outlets for the material. For instance, the raw material in group Cl does not differ significantly from material in group C5, however, the latter is a high ash fuel and can be used either as an initial raw material which would be a wasted resource when it could be used directly in combustion or for gasification. Coal wastes of the C2, C3 and C4 groups can be used in the manufacture of porous filters by the agglomeration method. However, material with the C3 parameter does not require the introduction of any additional fuel whereas C2 would require a reduction of fuel by adding very low organic carbon.material.

When burning in a rotary kiln it would be preferable to use material having Cl parameters since the other groups (C2-C5) would need a preliminary burning phase which adds appreciably to the cost of production. Likewise all the material in the Cl-C5 category could be used for the manufacture of ceramics but because of the fuel requirement of the process C3/C4/C5 material would be the most suitable.

Experience has shown that material type Cl and C2, depending on the levels of the B, D, and S parameters, can be used to produce wall tiles. For the production of facing tiles the only suitable material would be that with a Cla parameter.

Coal wastes C2 and C3 can be used as a raw material for most thermo-restoration processes. The degree of coalification of the organic matter (D) permits, for instance, the evaluation of the speed of caking during the production of the fillers and the quantity of organic gasses when heating up to 800-900 degrees celsius. Hence, for coal wastes with parameter D3 the vertical speed of caking could be from 5 to 10 mm/ minute, that is lower than for traditional clay materials , which are in the range 10 to 15 and 15 to 25 mm/minute. Alternatively material having the parameter Dl is not very suitable for heat treatment because at reduction conditions it is typical for the great bulk of fuel containing components to segregate into the gas phase at temperatures much lower than those necessary for the reduction of the silicon and aluminium compounds.

The sulphur content (S) of the waste material also determines the choice of direction for utilization.

TABLE 1.

SYSTEM OF INDEXING OF CLASSIFICATION GROUPS OF WASTE FROM COAL INDUSTRY

DESIGNATION OF PARAMETER	PARAMETER, GROUP, SUB-GROUP	INDEX NUMBER
A	SOURCE OF COAL WASTE	A
	(1) Coal mining waste :	A1
	(a) Opencast	A1a
	(b) Underground	A1b
	(c) Other	A1c
	(2) Coal preparation waste :	A2
	(a) Large dirt (13mm)	A2a
	(b) Medium sized dirt (13mm)	A2b
	(c) Fine dirt (1mm)	A2c
	(3) Heat treatment Waste :	A3
	(a) Ash/slag waste from combustion and gasification	A3a
	(b) Hydrogeneration waste (slurry)	A3b
	(4) Coal dump waste :	A4
	(a) Burnt	A4a
	(b) Unburnt	A4b
B	LITHOLOGICAL AND MINERALOGICAL CHARACTERISTICS	B
	(1) Clays :	B1
	(a) Kaolins	B1a
	(b) Hydromicas	B1b
	(c) Montmorillonites	B1c
	(2) Sandstones	B2
	(3) Carbonates :	B3
	(a) Calcitic dolomites	B3a
	(b) Siderites	B3b
C	CARBON CONTENT (ORGANIC)	C
	(1) Low carbon (0-4% C^{org})	C1
	(a) Less than 2% C^{org}	C1a
	(b) 2-4% C^{org}	C1b
	(2) Relatively low carbon (4-8% C^{org})	C2
	(3) Medium carbon (8-12% C^{org})	C3
	(4) Relatively high carbon (12-20% C^{org})	C4
	(5) High carbon (more than 20% C^{org})	C5
D	DEGREE OF COALIFICATION OF ORGANIC MASS	D
	(1) Low metamorphosis (C^f 75%) R^o = 0.49	D1
	(2) Medium metamorphosis (C^f=75-90%) R = 0.50 -2.49%	D2
	(3) High metamorphosis (C^f90%) R^o2.5%	D3
Fe	CONTENT OF IRON COMPOUNDS	Fe
	(1) Low iron (1.5% Fe_2O_3)	Fe1
	(2) Relatively low iron (1.5 -5% Fe_2O_3)	Fe2
	(3) Medium iron (5-12% Fe_2O_3)	Fe3
	(4) Relatively high iron (12-18% Fe_2O_3)	Fe4
	(5) High iron (18% Fe_2O_3)	Fe5
S	SULPHUR CONTENT	S
	(1) Low sulphur (0.5% S_t^d)	S1
	(2) Relatively low sulphur (0.5%-1.5% S_t^d)	S2
	(3) Relatively high sulphur (1.5-3% S_t^d)	S3
	(4) High sulphur (3% S_t^d)	S4

Coal wastes with the S1,S2 and S3 parameters are suitable for the manufacture of porous fillers; those with the S1 and S2 parameters can be used to produce fire resistant mate rials and wall tiles. Coal wastes of the S4 group can be used for in the lower layers of highway construction since they could benefit from the addition of various binders.

The above examples are merely a few of those that could have been discussed nad have bben chosen to demonstrate the influence the six main parameters have on identifying potential ways that coal mining wastes could be utilized. Ultimate choices will also depend on the technological process and equipment available.

REFERENCES

Shpirt, M.Ya & Ruban, V.A.
Utilization of mining operations and coal processes wastes in the USSR and the principles of their classification.
2nd Int. Sym. Reclamation, Treatment and Utilization of Coal Mining Wastes.
Ed. A.K.M.Rainbow, Nottingham University 1987.

Why Recycle?, Rainbow (ed.) © 1994 Balkema, Rotterdam, ISBN 90 5410 367 1

The main trends of utilization of coal mining wastes

A. K. M. Rainbow
MES International, Bedlington, UK

M. Ya. Shpirt
Institute of Fossil Fuels, Moscow, Russia

The amount of colliery spoil disposed of on colliery tips varies with the vend (ratio of coal to dirt) as well as on the level of coal production. In 1956 it was estimated that in the U.K about 25 million tonnes a year was deposited. When in the late 1970's the industry was producing 100 million tonnes of coal some 55 million tonnes were disposed onto tips. (5 million tonnes were disposed of in the sea). Due to the rapid closure of collieries in the last 10 years the amount of colliery spoil deposited has reduced to about the 25-30 million tonnes level.

In the United Kingdom the accumulation of colliery spoil over the years has resulted in an estimated 3,000 million tonnes of spoil owned by British Coal, in addition there are probably some 100 million tonnes not owned by British Coal.

It has been estimated that prior to 1965 most colliery spoil heaps were burning or had burnt out. Since 1965 the balance has been adjusted so that today an insignificant number of tips are burning and as a consequence the majority of tips are composed of un-burnt colliery spoil. This is because legislation introduced into the U.K. coal industry prohibited any newly constructed spoil tips to be allowed to burn.

For many years uses have been found for substantial tonnages of colliery spoil. For instance in the early 1960's, 5 million tonnes a year of burnt colliery spoil was used as common fill in the construction of road and railway embankments and the infilling of derelict or semi-derelict sites for building or other purposes. At the height of the motorway building programme in Britain as much as 10 million tonnes a year was being utilized.

For such applications burnt spoils were found, taking into account costs, to compare satisfactorily with naturally occuring rocks and engineering soils and so were often preferred when supplies were conveniently close to the point of use. In connection with the selection of engineering fills the major cost factor will always be transportation costs.

In addition to common fill "well-burnt non-plastic colliery spoil" meets the specification as a granular sub-base material in roadwork construction. However, such colliery spoils are only allowed to be used within the surface of the road if it is classified as "frost resistant" as determined by the frost susceptibility test. Many well-burnt non-plastic colliery spoils satisfy the frost susceptibility criteria, however, some fail this test.

Although the latter type of application made better use of the material, in proportion, this usage represents an insignificant tonnage when compared with that used as bulk fill.

Well -burnt colliery spoils have been used as under-floor fill for building - however - it is imperative that good control is exercised so that there is no risk of deterioration of concrete or mortar due to the migration of sulphates which may occur in well-burnt colliery spoils.

Minor applications for which well-burnt colliery spoils may be used are for tennis courts, sports tracks including motor sport tracks, though in some cases the material may require screening and/or crushing.

Well-burnt colliery spoil can only be used as an aggregate for making concrete building blocks-however- the soluble sulphate content might be the restrictive parameter.

Well burnt colliery spoil can only be produced if the bulk of the material could be prone to heating and susequent burning and because in the United Kingdom tips provide such material it was considered that unburnt colliery spoil would display the tendency to be liable to spontaneous heating and would prove to be an unsuitable material for construction projects. Because the supply of well-burnt colliery spoil was diminishing at a dramatic rate attention was focused on the potential utilization of unburnt colliery spoil.

In order to promote the use of unburnt colliery spoil it had to be demonstrated that when properly "engineered" unburnt spoil was not liable to spontaneous heating. Since it is an engineering requirement and therefore a specification requirement that material used in bulk earthworks should be well compacted it was possible to demonstrate that when the air voids ratio was less than 10% it was not possible for air to penetrate the mass of unburnt colliery spoil to promote any heating.

Another concern expressed by some engineers was that unburnt colliery spoil would breakdown in wet weather especially when being subjected to regular construction traffic. Contra to the fears of breakdown many embankments have been constructed in wet weather without serious delay.

A further important aspect was soon discovered concerning the materials' suitability - it was found under rigorous tests that unburnt colliery spoil was non-frost susceptible which is an important consideration in construction.

Generally, the material used will be taken from existing spoil heaps in preference to fresh material.
There are a number of reasons for this preference :-

(i) Fresh material will have a higher moisture content.
(ii) Spoil heap material is more likely to have stabilized both chemically and physically due to exposure to the elements.
(iii) The supply of material from the tip will be continuous whereas supply from the bunker could be erratic.

Unburnt colliery spoil is now a recognised secondary aggregate which can be used in highway and rail embankments, river and flood banks, especially in coal mining areas where there is a need for flood banks resulting from the general lowering of land due to mining subsidence lowering flood plains. Unburnt colliery spoil has also been used as fill in the construction of reservoirs, dams, ash lagoons and tailings ponds.

In all the previous mentioned applications the spoil, whether burnt or unburnt, has been used as an inert material, chemically unchanged by the application. Unburnt spoil can be used as a raw material for brick making and the manufacture of cement. In the United Kingdom it is estimated that 0.5 million tonnes of unburnt colliery spoil is used in brick making and 0.1 million tonnes used in the manufacture of cement per annum. The figures in regard to the level of brick making needs further explanation since in the 'hay-day' of coal production in Britain nearly every colliery supported a brick works and indeed the then National Coal Board was the countries largest brick producer. Bricks made from colliery spoil suffered from being variable in colour, texture and shape. Architects of the day preferred a more even colour, texture and shape and therefore sand faced bricks took over from the 'commons' produced at collieries; as a consequence NCB bricks ceased production. There is a trend today to revert to the 'antique' brick to the extent that a new (£11m) brickworks has been constructed at Tilmanstone in Kent to use colliery spoils as the feedstock.

In the mid 1960's the then NCB started a new technical and economic investigation of the possibilities of making lightweight aggregate or other aggregates from colliery spoil. There are two basic ways in which a suitable feed material can be satisfactorily converted into lightweight aggregate. One is to heat process it in lumps or pellet form, so that it "bloats". The other approach is to crush the feed to less than 5mm and to agglomerate the crushed material, leaving voids between the grains, and to sinter them sufficiently strongly to one another. The lightweight properties are derived mainly from the voids between the grains. Experience has shown that the sintering process is the most succesful of the two methods. The system depends upon a down-draught travelling grate on which a bed of agglomerated, crushed, fuel-containing feed is formed. The bed is ignited at the top, and as it travels forward the burning zone moves down the bed, indurating and sintering it into a 'cake' which after cooling is crushed and graded as required.

Experience has also shown that spoils from

bituminous coal fields give rise to an unacceptable effluent, however, the spoils from anthracite coal seams produce an excellent aggregate without effluent problems.

Engineering soils, of which unburnt colliery spoil can be considered to form a part, can be significantly improved by the addition of binders. The binders which have been used to enhance the performance of colliery spoils have included, bitumen, lime, pfa and cement - by far the most succesful has been the cement. For the cement bound colliery spoil to be economical an addition of no more than 10% cement is required. In many cases the addition of 5% has been sufficient to produce acceptable strengths. Cement bound colliery spoil can be used as sub-base material under roads or as hardstanding and/or pavement areas for workshops and storage.

REFERENCES

Rainbow.A.K.M A review of the state of the Art in Respect to the Risk of Spontaneous Combustion when using Minestone in Construction Works in the United KIngdom. United Nations ECE, Tatabanya, Hungary 1983.

Rainbow.A.K.M, Kettle.R.J. Cement Stabilized MInestone. United Nations ECE Tatabanya, Hungary 1983.

Rainbow.A.K.M. Minestone - it's use as a Secondary Aggregate. Sem. on Secondary Aggregates. Cwmbrau South Wales 1979.

Rainbow.A.K.M. The Utilization of waste from Coal Mining in United Kingdom. Bogazici Univ. Istanbul, Turkey 1982.

Why Recycle?, Rainbow (ed.) © 1994 Balkema, Rotterdam, ISBN 90 5410 367 1

Foundry sand recycling with the Clean Sand Cycle

Trevor J. Keirle
Keirle Engineering and Consultants, Lichfield, Staffs., UK

ABSTRACT: Economic management of the costs associated with the use of sand is now an essential part of efficient foundry practice. Modern standards demand higher precision in castings which has led to the widespread use of chemically bonded sands containing phenolic urethanes and esters. These chemicals and other contaminants picked up in the casting process have made sand re-use essential and the landfilling of this material is now environmentally unacceptable. Even when less hazardous sands can be landfilled, every ton of sand dumped has to be replaced by new quarried sand to maintain the mass balance of the foundry sand system. Keirle Engineering and Consultants (KEAC), have pioneered two thermal processes which enable foundry sand to be recycled and convert the fines and metal contaminated fractions into lightweight aggregates suitable for civil engineering use.

MECHANICAL RECLAMATION

There are many suppliers of mechanical sand reclamation systems and associated equipment. The majority clean the sand by the physical scrubbing action of sand upon sand - by vibration, by impacting the sand upon target plates, or by rotary or pneumatic propulsion of the sand. The type of plant best suited to the removal of the coatings that have been applied to the sand grains in the binder system should be chosen carefully either by historical preference or comparative assessment and analysis.

The criterion that should be applied is that the recovery of the maximum amount of sand required for moulds be achieved at the lowest acceptable and controllable sand quality. The mechanical recovery plant can be chosen from the range of vibratory and rotary designs available or be recommended and supplied by KEAC.

These criteria apply equally to both clay and chemically bonded sands as the mechanical reclamation system is equally applicable and is usually the lowest capital cost option.

This recovered sand is incapable of being returned to the moulding line when blended with new sand to give the required mechanical properties for casting. Unfortunately the sand quality is not good enough to enable re-bonding for coremaking

At this point then it is necessary to dispose of the residue of the sand to dump and purchase new sand for core making and blending.

The cost of new sand is not likely to decrease and the cost of dumping waste sand will almost certainly increase or be prohibited by environmental legislation in the case of chemically bonded sands.

As further economies and quality improvements are sought together with the wider use of chemically bonded sands then it has become less and less acceptable or desirable to dump used sand and the pressure continues to increase for the exclusion of leachates from landfill sites.

THERMAL RECLAMATION

The process of thermal reclamation has been addressed by KEAC using our fluidized bed technology while other manufacturers have chosen a number of alternative methods of heating the sand to thermal treatment temperatures.

KEAC can offer a modern steel lined rotary kiln of trefoil shape that will operate at temperatures up to 1200° C but it is generally accepted that the fluidized bed is a more efficient combustion device.

Around 1974 the first fluidized bed sand reclamation units were designed by the author's company

A fluidised-bed recycling plant designed by KEAC

Fluidfire and the units were licensed to Fordath. These were of a flat bed design based upon a flat ceramic tile for the distributor and the bed was fluidized using a premixed gas air combustion mixture and used the slump zone technology used in the construction of heat treatment furnaces for minimising expansions.

Fluidfire manufactured many of these beds and they were supplied by Fordath to a number of countries worldwide.

In 1982, on the formation of KEAC, the ceramic tile technology was superseded by a new all metal distributor design which overcame the problems of continual tile failure. This technique was evolved for the injection of a premixed gas - air fluidizing mixture into the bed using a matrix of elevated nozzles (photograph figure 4)using the natural insulation of bed material to allow cold low cost distributor designs for metal heat treatment furnaces using various pre-mixed controlled atmosphere conditions for processes such as carburising. Fluidized beds operating at 1250° C were supplied.

In the sand reclamation furnace the problem of poor thermal efficiency remained due to the restrictions on preheating the fluidizing air by the use of pre-mixed gas -air as the fluidizing medium. KEAC designed and has had patents granted for our multi -plenum direct injection system which uses concentric flows from stacked plenum chambers to elevated nozzles - incorporating the previous designs of slumped bed insulation to allow low cost construction of stacked plenum chambers - so that the combustion air may be preheated to a temperature close to the auto ignition temperature and the second plenum enabling the correct, controlled combustion of the fume from the sand preheat stage without the need for separate afterburning.

The waste heat in the system arises at two stages - in the hot gas exhaust which is recovered to combustion air preheat and in the discharged hot sand which is recovered to incoming sand preheat and fluidizing air preheat which carries into the bed the fume from the sand preheater.

The effects of these stages of heat recovery on specific fuel consumption are shown in Fig 2. and

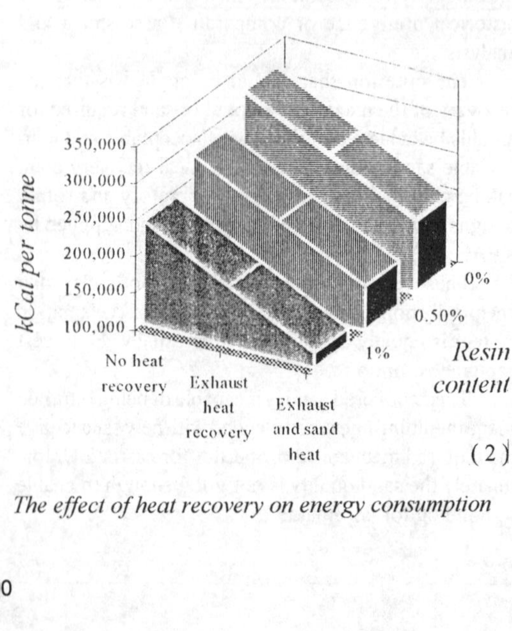

The effect of heat recovery on energy consumption

variations on this cycle have been used by the subsequent licensees of KEAC

Separate patents have been granted in all major industrial countries for the treatment of fine particles in a fluidized bed.

From 1988 the value and success of this patented technology was recognised and taken up rapidly by ECO GmbH in Germany, Dependable Foundry Eqt Co. Inc. in the USA and Maverex Ltd in UK. KEAC designed complex equipment with extremely good thermal efficiencies which was built by all three companies.

The first plant of these modern designs was installed in the Czech Republic at CZ Strakonice by ECO GmbH and had three individual fluidized beds in two separate enclosures. There followed quickly a number of other installations by Dependable and Maverex and the latter, in conjunction with Mechatherm Ltd., made a notable simplified installation at Rover Cars - Longbridge.

This was later followed by an new integrated design by KEAC of the more complex system with all three beds in a common enclosure, a plant being installed by Maverex Ltd in the Czech Republic at Skoda Cars.

After the closure of Maverex, KEAC supplied two basic units to the latest KEAC designs to Borden Boniface in France and this technology is now being used by Borden Boniface to service the French market and also by Mechatherm in the UK.

Recent sales in 1993 have been made by Dependable Foundry Equipment Co. with a 2.5 tph unit being sold into Australia and a smaller unit supplied in the USA.

A full list of reference plants is shown in Fig 3 and a complete range of modern fluidized bed furnaces for Foundry Sand Thermal Reclamation and Metal Heat Treatment are offered directly by KEAC using our own technology and performance and environmental data is available from many of the later installations.

ALL THE FOLLOWING PLANT ARE TO KEIRLE / KEAC DESIGNS AND THOSE SUPPLIED BY LICENSEES ARE INDICATED

COMPANY	COUNTRY	No x Size	SUPPLIER	SUPPLIER'S TRADENAME	YEAR
Doloro Stellite	UK	1 x 1.25 tph	Fluidfire - Fordath	Fluidfire	1978
Bevan Simpson	UK	2 x 1.25 tph	Fluidfire - Fordath	Fluidfire	1982/86
Hattersley	UK	1 x 1.25 tph	Fluidfire - Fordath	Fluidfire	1882
Aspera	Italy	2 x 1.25 tph	Fluidfire - Fordath	Fluidfire	1976
Borgo s Giovanni	Italy	3 x 1.25 tph	Fluidfire - Fordath	Fluidfire	1980
Moscow	Russia	7 x 1.25 tph	Fluidfire - Fordath	Fluidfire	1980/82
Leningrad	Russia	4 x 1.25 tph	Fluidfire - Fordath	Fluidfire	1982
Foundry	Iraq	1 x 5 tph	Fluidfire - Fordath	Fluidfire	1983
Musso	Italy	1 x 1.25 tph	Fluidfire - Fordath	Fluidfire	1985
CZ Strakonice	CSFR	1 x 1.25 tph	KEAC - ECO	Ecofire	1987 *
Wellman Dynamics	USA	1 x 1.25 tph	KEAC - Dependable	Calcifire	1988
Rover Cars	UK	2 x 4 tph	KEAC - Maverex	Thermocon/Rototherm	1989*
Armstrong Mould	USA	1 x 0.25 tph	KEAC - Dependable	Calcifire	1990 *
Bruel - West Yorks	UK	1 x 4 tph	KEAC - Maverex	Thermocon/Rototherm	1991*
China Railways	China	1 x1.25 tph	KEAC - Maverex	Thermocon/Rototherm	1991*
Skoda Cars	CSFR	1 x 1.25 tph	KEAC - Maverex	Thermocon/Rototherm	1991*
Zeus Aluminium	UK	1 x 0.5 tph	KEAC - Maverex	Thermocon/Rototherm	1992*
Borden Boniface	France	1 x 0.5 tph	KEAC	KSR500	1992**
Marly	France	1 x 0.5 tph	KEAC - Borden	Regetherm	1992**
Warman Int	Australia	1 x 2.5 tph	KEAC - Dependable	Calcifire	1993 *
Cobart Refract	USA	1 x 0.5 tph	KEAC - Dependable	Calcifire	1993 *
Sofogir	France	1 x 0.5 tph	KEAC - Borden	Regetherm	1993

* Licensee Manufacture ** KEAC Manufacture (3)

CHEMICALLY BONDED SANDS

The problems associated with the reclamation of chemically bonded sands are well understood and there are many manufacturers offering equipment to recycle sand from those processes. The residues from a well designed chemically bonded sand system are confined to a small amount of waste sand and some fine material from the dust collectors - the treatment of fine particles in a fluidized bed also being the subject of KEAC patents.

The use of thermal reclamation allows the sand to be cleaned and recycled to a condition and contaminant specification that is equivalent to or better than new sand.

In economic terms this means that new sand purchases are reduced and because the new sand is recovered from the waste sand the dumping charges are reduced.

The cost of treatment and the capital plant required has been well established for a number of years and as the disposal prices increase then the case for thermal reclamation improves. A recent study for a Czech foundry in Brno showed that a KEAC thermal reclamation furnace on Furan sand would attract only a 1.5 year inclusive payback.

Using thermal reclamation to achieve new sand quality allows cores to be made from reclaimed sand.

The reclamation of chemically bonded sands is well documented and understood, the easiest and most documented being the cold box and metal mould processes used in the production of aluminium automotive engine components.

The resin - binder systems are being improved continually to allow better mechanical and thermal reclamation of the sands. Chemical additives are now available to overcome some of the fluidization problems associate with some processes. These problems of bed fluidization can occur when the thermal reclaimer temperature is limited by the chemical compositions of the resin binder systems to a temperature that is lower than that required by the legal requirements on afterburning.

The fluidized bed has some dispensations in its role as an incinerator if the freeboard above the bed is correctly designed and operated and KEAC specialises in chemical waste disposal, recycling and recovery in partnership with the UK electricity industry research company EA Technology Ltd at Capenhurst England where it has a demonstration fluidized bed facility that operates up to 1200° C to meet chlorinated hydrocarbon incineration requirements.

The photograph illustrates the inside of a KEAC designed furnace showing the elevated patented nozzle design to introduce a gas/air mixture

CLAY BONDED SANDS

The thermal reclamation of green sand to core sand quality has placed unresolved demands on existing technology and development resources for a number of years.

Since the KEAC license was granted to Dependable Foundry Equipment Co. Inc. USA - it has clearly been proven that the progress has been made in this difficult direction to enable the recovery of green sand to core making quality and the process patented.

The re-bonding quality and the manufacture of satisfactory cores from thermally reclaimed green sand has been demonstrated to all the major USA automotive manufacturers and a paper was presented to The Instate of British Foundrymen at its recent seminar at CTI Sheffield, England on this aspect of reclamation by the president of the KEAC licensee in the USA - Dependable Foundry Eqt Co. Inc. - who have patented the process used.

The result of successful mechanical and thermal reclamation is that there is very little new sand purchase required and the costs of dumping are significantly reduced.

AGGREGATION

Assuming now that the thermal reclamation process has met the demand for the majority of the new sand then the remaining residue of sand going to disposal costs of fines from the dust collection plants which require compacion or agglomeration, sludges containing metallic dusts from the fettling shop if wet arresters are used and sand losses from the casting floor.

The KEAC steel trefoil kiln will convert these materials into a stable non-leaching building aggregate material at a loose bulk density of 500 - 600 kg / cu m that can be used in concrete, lightweight blocks or base fill.

The aggregation of clays and sludges is part of the KEAC technology used in the treatment of industrial and sewage sludges in conjunction with a major UK owned international construction company. Projects for this treatment are in Hong Kong Taiwan and Mainland China as well as the UK.

From a foundry the largest volumes of sand residue generally arise from the green sand process which is usually a larger consumer of new sand. The residues of sand which cannot be treated by reclamation processes must be dumped - at not insignificant cost.

This disposal cost detracts from the profitability of a foundry but it can be turned into an income by converting this waste sand and fine material into a building aggregate at the same time stabilising the products into a no leaching material. Before the aggregation process is considered the maximum amount of sand for re-use should be recovered.

The size of aggregation plant required for on-line treatment of waste sand from a complete foundry sand reclamation system is quite small but in the case of a foundry having accumulated a large "on site" dump of waste sand then the conversion of this waste sand dump into useful building materials can be undertaken as a land remediation project to meet environmental requirements which may attract financial grant aid. At the same time income can be generated from the sale of aggregate, the value being dependant upon the geographical location of the foundry.

The ultimate aim of this composite technology is that a foundry need not be a contributor to environmental pollution from its sand. In the process of reducing the waste output a useful product is created and the resultant sales of aggregate may lead to a profitable secondary business.

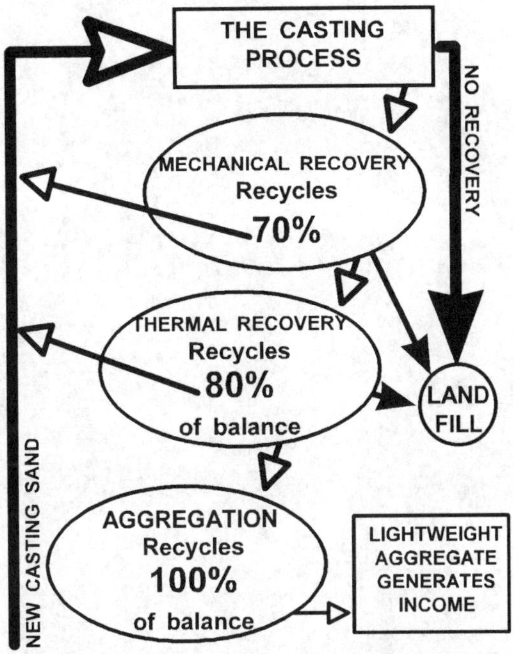

(5)

With the CLEAN SAND CYCLE all recoverable sand is re-used and the finest particles recycled as lightweight aggregates. Nothing is landfilled.

The plant can be very cost effective and short payback times achieved dependant upon disposal charges and tonnages.

The process can equally be applied to a whole range of contaminated and toxic waste materials - including contaminated land that requires remediation - and the resulting granular products are readily usable and saleable into the aggregate market.

This technology is was first established in 1982 for colliery washery waste and samples can be made from customer residues in a laboratory kiln.

KEAC is committed to the continuing development of these technologies throughout the world with wholly owned companies, associate licensees and joint ventures. Companies and individuals requiring further information are invited to contact the author.

Why Recycle?, Rainbow (ed.) © 1994 Balkema, Rotterdam, ISBN 90 5410 367 1

The *Modern Alchemy*™ process for recycling metals from chemical by-products

Trevor J. Keirle & Chris Cutchey
K.C. Process Limited, Lichfield, Staffs., UK

ABSTRACT: Hundreds of thousands of tonnes of metals are used each year in chemical processes and industrial manufacturing of which only a tiny proportion is recycled. In the production of food, detergents, perfumes, and chemical intermediates, metals such as nickel, copper, platinum and molybdenum are used as catalysts and often incinerated and dumped or simply thrown away after use. *Modern Alchemy*™ is a recently patented process using advanced high temperature fluidised-bed techniques to recover fine particulate metals entrained in filtrate or other chemical by-product streams. It can be made small enough to process the by-product on-stream at the "end-of-pipe" within the production works, or larger plants can be established as stand alone recycling facilities. It is, above all, environmentally-friendly and, coupled with downstream chemical or electro-chemical conversion processes - as pioneered by EA Technology Limited at Capenhurst, will ensure that all the metals and particulates in these residues are returned to a completely usable product such as copper sulphate, nickel chloride, or metal foil.

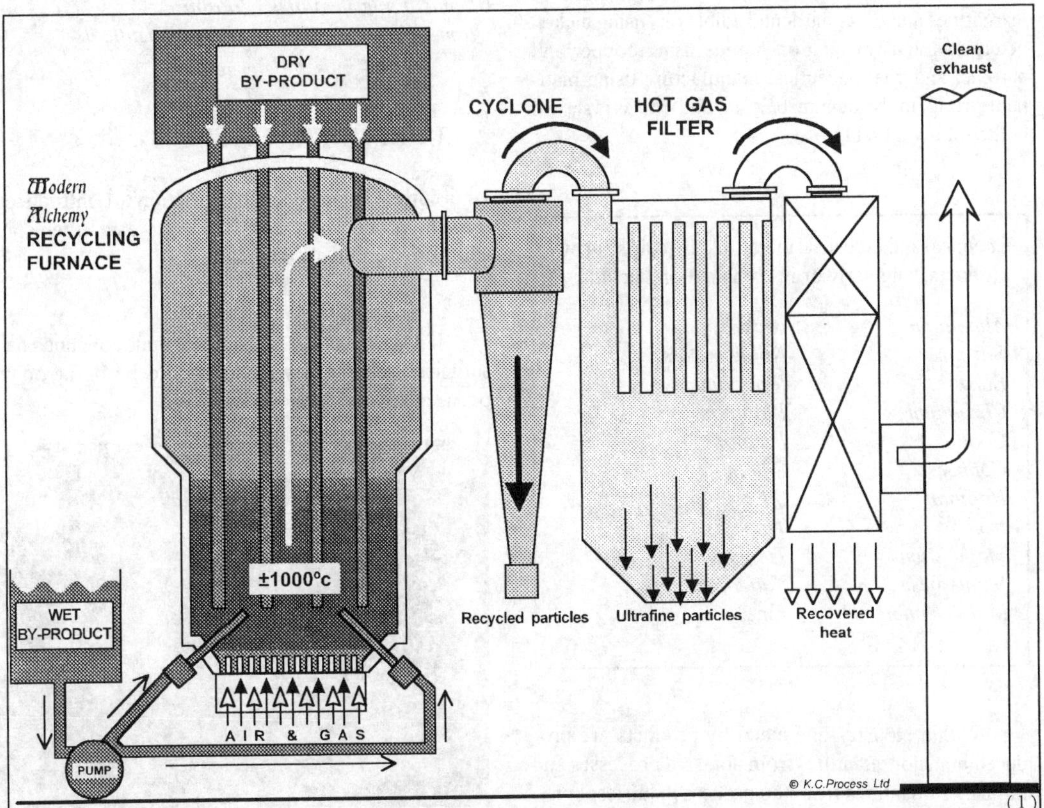

1

Spent catalyst

1. THE ORIGINS OF CHEMICAL AND INDUSTRIAL METAL BY-PRODUCTS

One of the principal uses of metals in chemical processes is as a catalyst to bring about a chemical change of reaction in another substance. The actual metal does not itself undergo a permanent change and may be re-used or recycled.

Typical processes include the manufacture of margarines and other hardened edible oils using nickel, detergents and washing-up liquids using copper and chrome, and pharmaceutical manufacture using platinum-group, molybdenum, barium, and iron as is shown in the following table:-

Chemical and industrial metals likely to arise in by-products, singularly or in combination, include:

Aluminium	*Nickel*
Barium	*Noble metals*
Boron	*Paladium*
Chromium	*Platinum*
Cobalt	*Precious*
Copper	*Silicon*
Hafnium	*Silver*
Iron	*Tin*
Magnesium	*Tungsten*
Manganese	*Vanadium*
Molybdenum	Zinc

(3)

Within industry fine metal by-products are produced in major quantities from abrasive processes and in smaller quantities from the process indutries.

Types of processes using metal-bearing catalysts:

Alcoholysis	*Emission control*
Alkylation	*Esterification*
Amination	*Hydrogenation*
Ammonia synthesis	*Hydrogenolysis*
Aromatics production	*Isomerisation*
Cracking	*Methanation*
Dehydrogenation	*Oxidation*
Desulphurisation	*Polymerisation*

Uranium fuel reprocessing

(4)

Metal-bearing residues and by-products may take the form of slurries, sludges, grindings, filter cake, or muds. They may be very wet, oily, sticky, and contain organics, inorganics, poisons, and various hazardous materials. They usually contain volatiles which add to the problems for recycling

Some industries where catalysts are used:

Agricultural chemicals	*Food processing*
Agrocultural chemicals	*Nuclear*
Chemical intermediates	*Paints*
Cosmetics	*Pharmaceuticals*

Soaps & detergents (5)

2. CURRENT PRACTICE

Until current environmental legislation was introduced these by-products were treated as a waste, often consigned to uncontrolled landfill.

2.1 Smokestack

If it was considered that the metals contained had sufficient value, the volatiles would be burnt off in ovens of rotary kilns:-

Both the company photographed in (6) and the works shown below had credible processes to recover metals from spent catalysts.

Both recycling operations have now been closed down as a result of environmental regulation

The black plume emanating from the chimney is fine nickel dust which is converting itself to oxide and being deposited on the adjoining fields and watercourse

There is now only one substantial thermal recycling plant left operating in the western world. This is in Austria

2.2 Chemical dissolution & refining

Although many metals have been, and some still are, recycled by dissolving them in acids and converting to metal salts or back to metals using advanced electrochemical techniques, total metal recovery has not been generally possible.

Neither has it been possible to treat one by-product without creating another effluent which itself will pose a further disposal problem.

Although it has been possible to landfill these effluents after neutralising them, these routes are becoming more expensive and less environmentally acceptable making chemical recycling processes less viable and unattractive.

2.3 Incineration

Incineration undoubtedly destroys the volatile fraction of these by-products and modern fume control techniques render them relatively harmless.

However by their very nature commercial incinerators are designed to handle vast quantities of volatile materials and cannot have their production dedicated to treating materials on an individual by-product basis.

Thus the residual ash they produce contains a vast cocktail of metals and other inert solids from which any type of recycling is not realistically feasible.

As a result the ash is deposited into controlled landfill sites.

2.4 Landfill

The majority of metal-bearing residues are leachable and therefore pose a potential environmental problem to water courses, flora and fauna.

Even the ash from incineration, which is currently deposited in fully lined tips to prevent such contamination, may one day be disturbed and would then produce a massive environmental threat for future generations.

3. PARTICLE RECYCLING

3.1 Particle sizes

To work effectively catalysts must combine having the greatest surface area of active metal available for the required reaction to occur and to be able to disperse through the material being catalysed.

This condition is produced by making the particles as small as possible while at the same time not so small that they cannot be effectively filtered out after the reaction has occurred.

Typically a fresh catalyst of the "fatty" type may have a particle range to the following graph:-

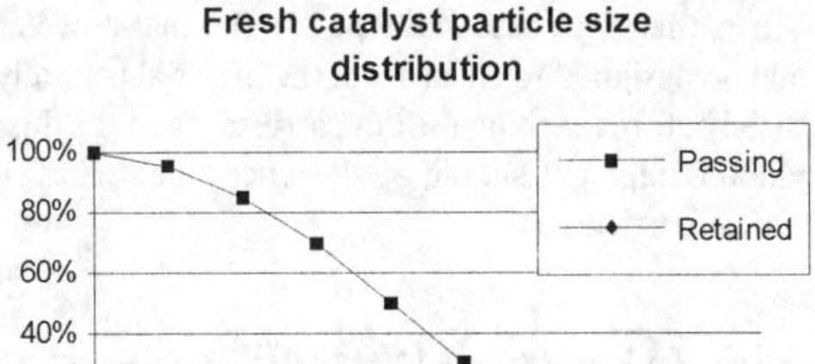

(8)

Whereas after use and recycling the particle size may reduce, perhaps due to oxidation and/or attrition:;

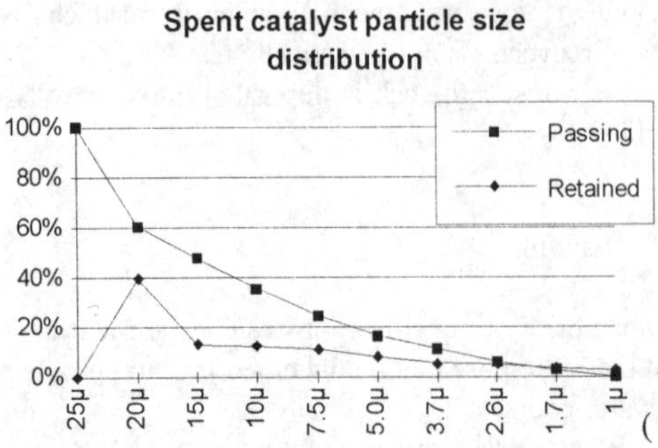

(9)

This size reduction is one of the properties of recycling catalysts and other fine particulates that has often gone unnoticed by the smokestack operator and which has led to loss of metal and neighbourhood contamination.

3.2 Oxidation

Many metals, such as nickel and iron, when in a very fine particle form will oxidise when exposed to oxygen.

This process, or reaction, is exothermic - that is the particles gain heat, in some cases quite considerable amounts, which is why the metal filters used by the plant in the photograph (7) were ineffective.

The plant designer had assumed that by drawing the particulates through a screen of water the nickel would cool down and could then be recovered as a sludge.

In fact the exothermic reaction was so great that despite being drawn through two such "filters" the still-heating particles were expelled out of the stack.

It is therefore essential that any thermal recycler should be designed to ensure that the material is totally oxidised before reaching the dust arresters and the dust handling equipment should be designed to catch even the finest particles.

4. The *Modern Alchemy*™ process

Modern Alchemy™ is a proven, high-technology state-of-the-art process demonstrating recycling at its best.

The entire process is self-contained from the instant that the spent catalyst - or whatever the feedstock happens to be - is injected or blown in to the fluidised-bed, until it is released as reclaimed fine particles and the organics or combustibles are consumed within the furnace.

The process can be designed for ''wet'' material (fig 10) which is pumped through the base of the furnace straight into the hot bed where it is instantaneously volatilised. The fine particles travel upwards through the hot bed material (usually alumina which remains in the furnace) emerging at the top totally stable and, if appropriate, oxidised:

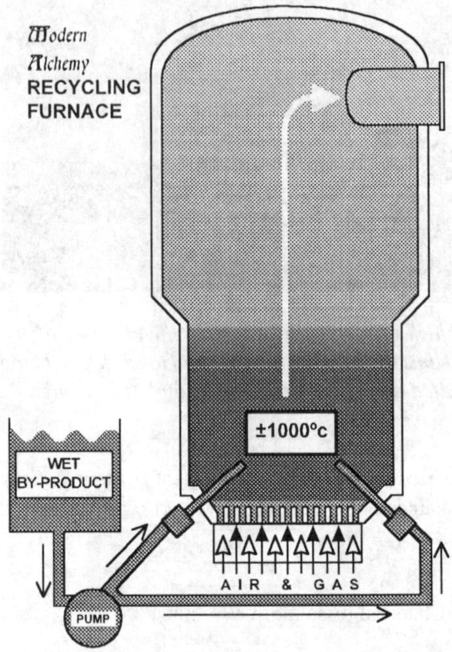

(10)

Some by-products come in the form of liquids, often with the solid fraction in suspension.

In this case they will be introduced to the fluidised bed by pumping through metered injectors with the flow rate controlled by a computerised programmable logic controller.

However if the materials are dry they will be blown into the heart of the bed from above (fig 11). Should they be sticky or otherwise fairly immobile, they may be mixed with product or other suitable material to form granules which can be blown in.

For maximum versatility however, the furnace may be designed to take wet or dry feed (fig 12).

This is particularly usefull for multi-purpose plants although it should be pointed out that campaign work on different base material could lead to high downtimes for internal cleaning to avoid cross-contamination of materials.

The calcined, totally dry, product is expelled

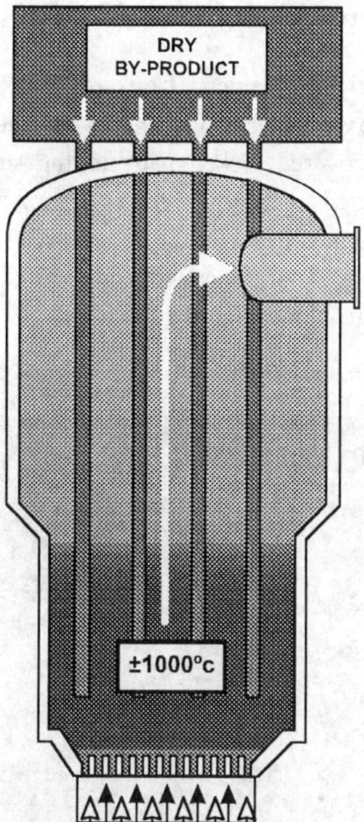

(11)

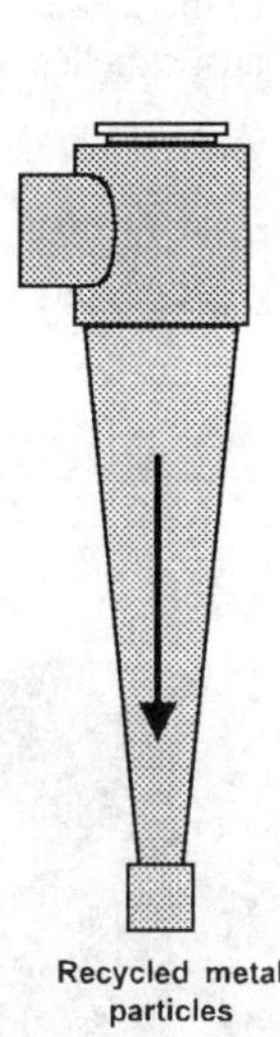

(13)

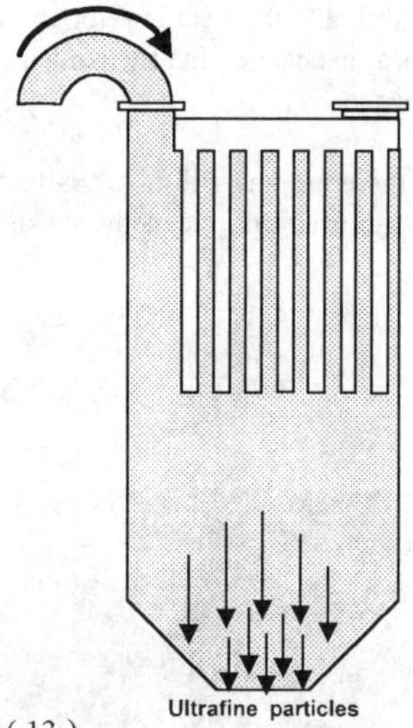

(14)

upwards out of the fluidised bed into a high-efficiency cyclone (13) which removes between 70% and 95% of the particles from the gas stream.

The amount captured varies according to the particle size, density, and gas flow. When recycling nickel-bearing catalysts particles of an average 7μm were captured at this stage.

The finest particles are passed over to a hot filter (fig 14), designed to operate at temperatures up to 1000°c, and removes 99.95% of all particles down to 1 μm. If finer particles are present further filtration or electrostatic precipitation will be necessary although this does not apply to most by-products.

Modern Alchemy
RECYCLING FURNACE

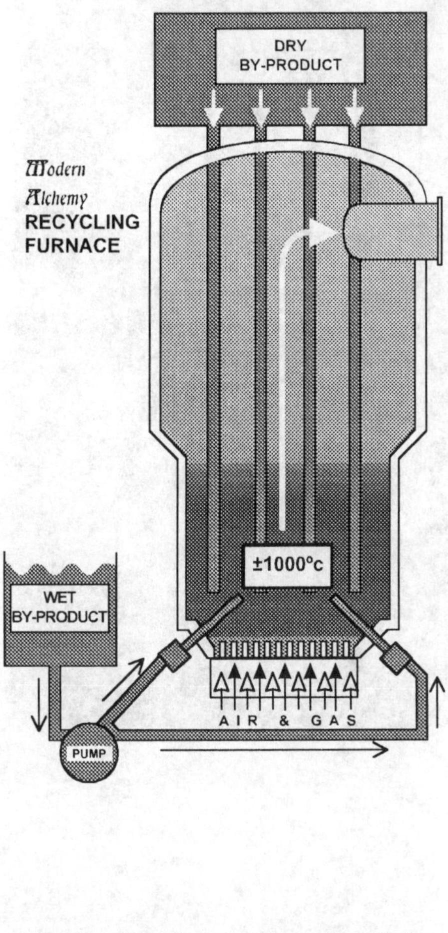

(12)

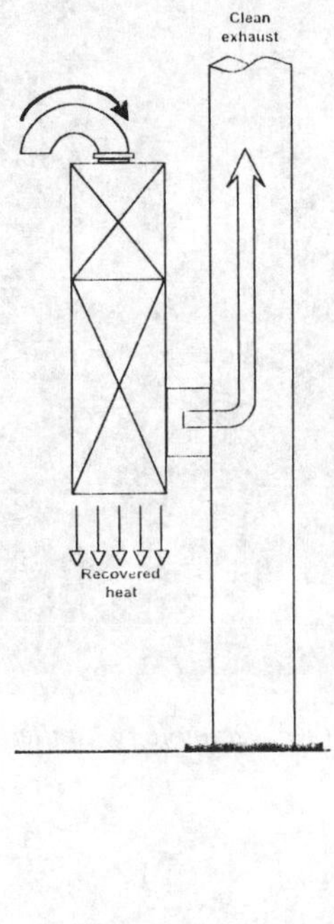

(15)

And finally the heat is removed from the gasses with a heat exchanger and the cooled exhaust expelled to atmosphere.

If there are any still-noxious fumes present they can be captured and cleaned by existing proprietary means.

5. CHEMICAL CONVERSION

The original plant, at Cookson Industrial Materials' Bootle works was established with a view to selling all the recovered nickel oxide to the steel industry in briquetted form. However it was recognised that for the maximum added value to be realised from the process the metal should be converted to chemical salts such as sulphate, chloride, or hydroxide.

The test rig at EA Technology Limited

An investigative programme into this was carried out by Cookson Group Technology Centres at Perivale and Oxford, and even more work has been undertaken by EA Technology Ltd (formerly the Electricity Research and Development Centre) at Capenhurst, near Chester, England.

EA Technology, who were in receipt of a European Grant for research in catalyst recycling, achieved 100% recovery from the nickel oxide produced in the *Modern Alchemy*™ plant in April 1992. Additional work is being undertaken to ''fine tune'' the process.

6. TEST FACILITY

It seemed natural that this cross-fertilisation of technologies - this ''synergy'' - between the two companies should be promoted so the old test rig was totally refurbished to take wet or dry feed materials and a hot filter system added. It was commissioned in Lab 9 at Capenhurst in April 1993.

EA Technology Limited is now the exclusive world testing centre for customers by-product evaluation and appraise samples under controlled and monitored operating conditions faithfully reproducing the *Modern Alchemy*™ process as it would in full-scale operation.

Their facilities include sample analysis and preparation, full analysis of product including particle size and gas stream evaluation, and full environmental impact assessment for operating to customer scale requirements.

A comprehensive report and recommendation process can be instigated for every stage of the trial and information obtained may be protected by binding customer confidentiality/secrecy agreements.

Why Recycle?, Rainbow (ed.) © 1994 Balkema, Rotterdam, ISBN 90 5410 367 1

Vertical floatation melting furnace

O.H. Perry
Stein Atkinson Stordy Ltd, Wolverhampton, UK

ABSTRACTS: The paper shows the present development of the Vertical Floatation Melter (VFM) and discloses the principles of operation, including the first two years of development from the laboratories to an operational plant.
The paper also traces the experience of the first twelve months of the operation of the furnace and considers further applications over the coming years.

INTRODUCTION:

Why develop the VFM ? The main driving force for the development was the lack of a reliable environmentally friendly method of Melting Decoated Used Beverage Can Shreds, (DUBCS).
The channel induction furnace can be used for melting DUBCS but the main disadvantage with an induction furnace is the process is not continuous and therefore it is difficult to hot connect. Also, generally in most countries electrical furnaces, because of capital costs and running costs, are made prohibitive.
The most common method of melting DUBCS is the Sidewell melter, either using a Vortex pump or a Puddler to mechanically submerge the shreds. This Sidewell has many disadvantages:-

1) Fluxing salts are used during the melting process, therefore the system produces Toxic black dross. The fluxing salt flows vary between 1-3% of production. Assuming only 1% salt is used on a 20,000 tonnes per year plant, the plant would produce 200 tonnes of toxic black dross per annum.
2) Reliability. The system needs to use immersed pumps and their life is very short.
3) Fuel consumption of a Vortex Melter is high; between 1500 and 2000 Btu's per lb.
4) Metal loss is high, between 3-5%.

Having looked at the alternatives we saw a great need for a better DUBCS Melting Furnace.

FACTORS FOR LOW METAL LOSS:

The first factor was to see what was needed to produce low metal loss as this is the key to economic operation of the furnace. The three main parameters for low metal loss are:

(1) Low Furnace Temperature (Figure 1) at 777°C, 1,430°F and above, the oxidation rate increases dramatically; therefore by melting the shreds between 700°C, 1,300°F to 750°C, 1,400°F, oxidation is low.

(2) Fast Melting (Figure 2). Oxidation is also time related as the longer time it takes during the latent heat of fusion stage, the higher the metal loss. This has far greater influence

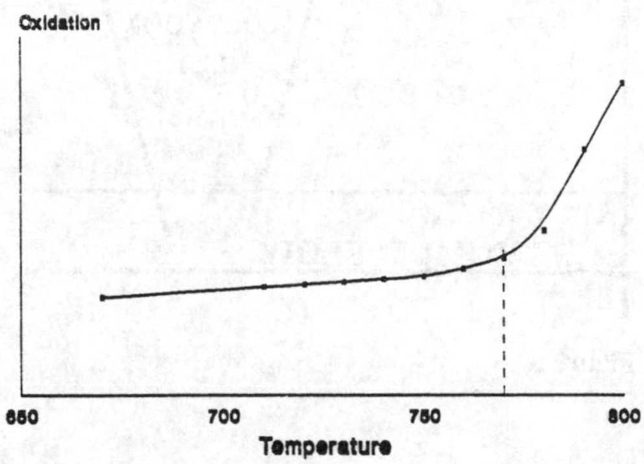

Figure 1 770°C Temperature

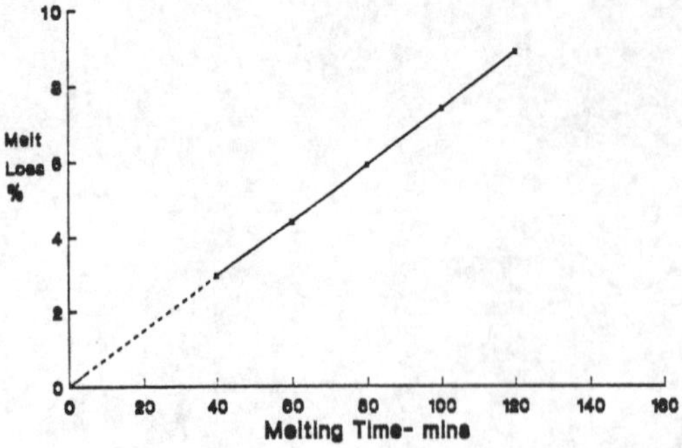

Figure 2

when surface to mass area is high, i.e. thin foil.

(3) Low Oxygen Atmosphere. Oxidation is also effected by the amount of oxygen in the atmosphere. This is why submersion is used in sidewell melters.

PRINCIPLES:

The idea of floating the DUBCS in a hot gas air stream is to give rapid melting at low temperatures. Figure 3 shows a cone cold test between zero velocity and NF when no DUBCS will float between NF and AF of shreds floated.

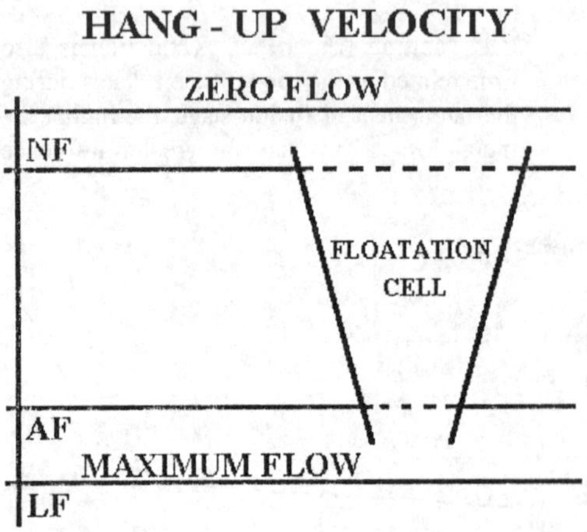

Figure 3

COLD TEST:

A 10ft high perspex model was built to test this at the University.

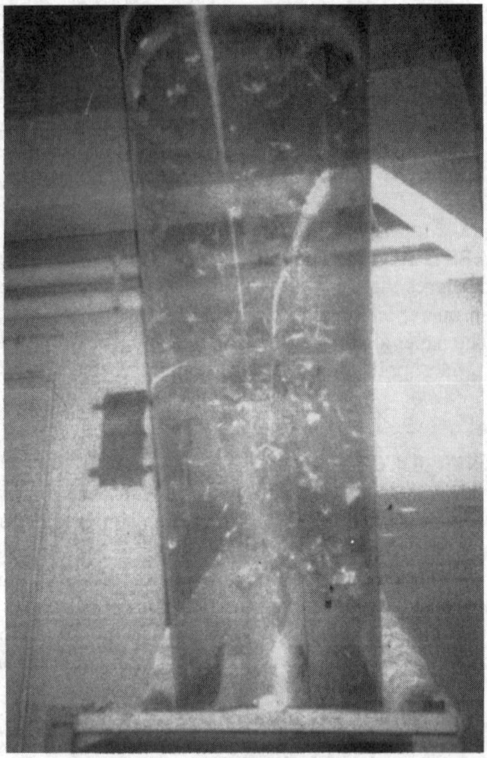

Figure 4

DUBCS were charged into the top of the cone and all floated. Changes in shredding grades did not have any measurable effect on floatation range. The standard test was if the shreds remained in the cone for more than four hours, the system was deemed stable.

HOT TESTS:

If the DUBCS melted into droplets, the velocity required to float them is greater than the velocity AF figure 3. We now have a solid liquid aluminum separator.

FURNACE DESIGN:

The hot shreds are charged into the cone, they float until they become liquid then the droplets fall into the bath below. In operation, most of the droplets agglomerate and run down the lower half of the cone wall then fall into the bath. The driving momentum of the system is the recirculation fan which blows recirculating gas into the holding chamber and then up the cone.
The recirculating gas enters the settling chamber where any carry over fines settle and melt running down the sloping floor and return to the cone.
The recirculating gas travels down a duct to the recirculating fan which blows recirculating gas into the holding chamber where the burner tops up the heat lost to the shreds.
This design meets the three needs of low metal loss:

(1) Melt at low temperature 700°C, 1,300°F, to 750°C, 1,400°F.
(2) Melt fast, 5 to 15 seconds.
(3) Melt with low 2/5% oxygen.

Figure 5

FULL PRODUCTION PLANT:

The Vertical Floatation Melter went into operation on the 14th September 1992.
An air transportation system is used to move the shreds from the Decoater to the VFM; the air transportation system is extremely flexible:-

From Decoater to VFM
From Decoater to Induction Melting Furnace
From Decoater to storage
From storage to VFM
From storage to Induction Melter
From storage to storage

The advantage of the air transport system is:

(1) The shreds travel fast from the decoater to the melter (small heat loss).
(2) No dust etc. enters the surrounding environment.
(3) Changing from VFM feed to induction melting feed can be instantaneous.
(4) It is easy to feed to and from the storage areas.

PRODUCTION TRIALS:

Melt loss, as previously stated, is the key to efficiency and it is most important to define how metal loss was measured.
It is accepted that an induction furnace has the lowest metal loss, so the induction furnace can be used as a benchmark for the metal loss trials.

FEED IN:

Because the air transportation system is used, it is possible to weigh the shreds leaving the decoater. Every hour, for a time period, the shreds would be diverted to the storage area where they were collected in a stage weigh bin.

These tests were carried out over a 14 days, 24 hour test period, i.e. 336 hours.

Also, every time the induction furnace required material, the air transportation system would feed it. The induction furnace had load cells fitted which gave us weight, feed into induction furnace, plus a further check on the feed rate.

Now having an accurate feed into the V.F.M, we now check output.
Output is in two sections:

(1) Metal Out
(2) Dross.

METAL OUT:

The ingot casts are weighed, plus any sows resulting from launder filter draining, minus any alloy additions.

Dross:

All dross was weighed and measured for aluminum recovery.
Therefore, Metal Loss equals:
Metal in - (Metal out + Recovery Metal in dross).
Metal loss over the 14 day period is given in Table 1 below:-

TABLE 1 METAL LOSS

	Weight	%
Metal In	840 tonnes	100
Metal Out	762 tonnes	90.7
Dross Weight	55 tonnes	6.5
Al in Dross	36 tonnes	65
Metal Loss	23 tonnes	2.7

FUEL CONSUMPTION:

It is determined by two factors on a V.F.M. throughput and DUBCS's temperature.
Throughput standing losses are fixed, therefore higher the throughput (closer to designed), lower the fuel consumption.

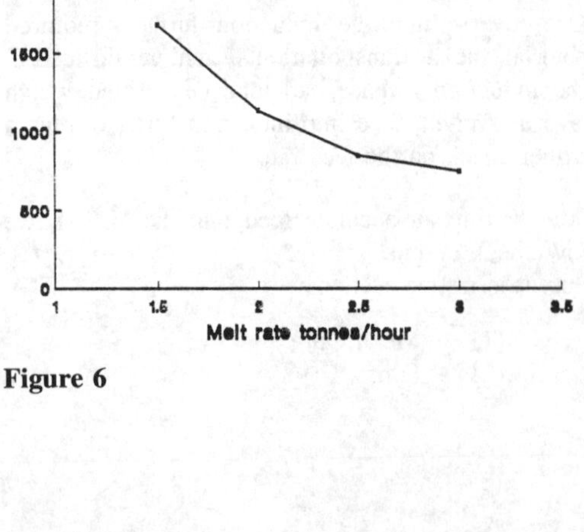

Figure 6

DECOATED USED BEVERAGE CAN SHREDS TEMPERATURE:

Hotter the DUBCS the less energy is required to heat them. Typical DUBCS temperature from decoating is 300ºC, 470ºF.

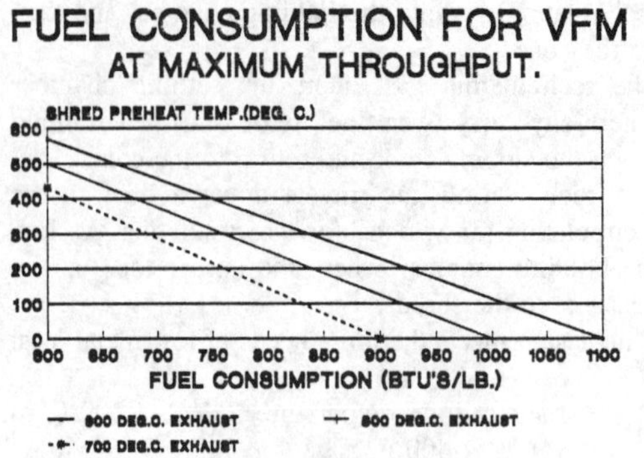

Figure 7

CONCLUSION:

The VFM has proved to be a reliable DUBCS Melter with fuel consumption figures that are well below other methods of gas fired furnaces.
We are now carrying out an intense two month metal loss trial where we believe the VFM will produce lower metal loss figures than stated in this paper.

FUTURE:

We have already melted closure stock, foil, scalping chippings etc. Some of these products could only have been melted in an Induction Furnace. The VFM has proved to be a reliable and cost effective solution to melting shredded, thin gauge material.

Why Recycle?, Rainbow (ed.) © 1994 Balkema, Rotterdam, ISBN 90 5410 367 1

Recycling of domestic refuse

P.R.E.Welby-Everard
SSI Schaefer Limited, Basingstoke, UK

ABSTRACT: The recycling of domestic refuse is now receiving increasing attention by local authorities in the United Kingdom. This paper is an overview of the methods currently employed and undergoing trials.

1 INTRODUCTION

During the last ten years, the requirement to recycle the largest possible quantity of our domestic refuse has increasingly come to the forefront of our consciousness as part of the general rise in environmental awareness.

In Germany there is a highly developed system of recycling, largely based on the organisation known as DSD or Duales System Deutschland, which effectively collects large quantities of separated material. At present thou there are still some problems to solve concerning finance and reprocessing.

In England there is a Government Directive to the effect that 25% of the total tonnage of domestic refuse must be recycled by the turn of the century.

2 SETTING THE OBJECTIVES

Before embarking on any recycling scheme it is necessary to define the objective to be achieved. It may be that all that is required is to achieve the Governments Directive of 25% of total refuse recycled by the end of this century at minimum cost. It may however, be to divert the maximum amount of material from the waste stream in order to save precious and expensive landfill space. It may also be to save the maximum amount of valuable raw materials and energy, or it may be (and often is) a combination of all these objectives.

It is this objective when taken into account that the finance available and the markets available for the recycled material which will determine the system or systems to be employed.

3 THE OPTIONS

A) The Bring System

This method involves the setting up of centralised amenity recycling centres to which the public brings the material. The containers used in these centres vary from the small 1100 litre wheeled bins to the large rolonoff containers which are used on the very large sites.

This system is used by many council's throughout the United Kingdom in various applications from the small mini recycling centre within individual housing estates or villages, to the very large urban amenity sites.

There are various container types in use including the 1100 litre wheeled bins, the bottom opening igloos of various types, the 3.2 cbm Grumbach containers which can be divided into 2 or 3 separate compartments as well as skips and the very large rolonoff containers.

The advantages of the Bring System are that it is simple to install and comparatively low in cost.

The disadvantges are that there may be difficulties in finding suitable sites for the containers, that these sites may not be available to many people who are not mobile and that the quantity of material collected in them will be comparatively small. This will vary according to the density of the sites but even with a density of one site per 500 - 1000 head of population it is unusual to obtain much more than 20% of the total refuse tonnage.

B) Segregation and Collection at Source

By this method the householder separates the recyclable material from the residual refuse and possibly into its component fractions. It is then collected and either sent direct for reprocessing or sorted into its component fractions in a sorting plant and then sent for reprocessing depending on the collection method employed.

The advantages of this method are that it collects a much greater percentage of the available material (upto 90%).

The disadvantages are that the cost is considerably higher thou this may vary in the future depending on disposal costs and the value obtained for the recycled material.

The main methods in use for collection at source are:

1) The Plastic Box Scheme

This scheme orginated in America and Canada. It involves a number of plastic boxes issued to each house. The householder then separates the recyclable materials into each box. The boxes are then placed at the front of the house and are then emptied by hand into a vehicle which is divided into compartments.

The advantages of this method are that it gives clean separation of the material fractions and it gives a good service. The capital cost of installing it is also comparatively low.

The disadvantages are that operating costs are very high, since it is very labour intensive and involves a separate fleet of vehicles as well as a separate labour force, in addition to those used to carry out the normal refuse collection service.

2) Wheeled Bins

These are used in two basic forms:

a) The split wheeled bin

In this system the wheeled bin is divided into two segments, each of which holds a different fraction. Both fractions are then emptied into a divided truck.

The advantages are that there is only one bin per property which is emptied every week.

The disadvantages are that the conversion of the bin is comparatively expensive, the vehicle is also expensive and complicated, the bin is not colour coded and experience indicates that contamination of the refuse fractions is very likely to occur due to the wrong items being put in the wrong compartment in the bin. Also, due to the fact that the vehicle contains two separate fractions of refuse, it is unlikely to be fully loaded when one compartment is full and the vehicle has to go to tip. It also has to be discharged at two separate locations and efficient utilisation of the vehicle is therefore unlikely to occur causing a loss of production efficiency on the collection round.

It is for these reasons, that SSI Schaefer have decided not to include this system in their range.

b) Dual or Multi Bin

This is the system which SSI Schaefer offer as part of their range of systems. With this system each property has two (or sometimes more) wheeled bins of different colours. Into each bin goes a different refuse fraction and the bins are then emptied in rotation by the the same vehicle and crew. The refuse is then either returned directly for reprocessing or if mix dry recyclables are being collected in the second bin through a separation plant (MRF) and then for reprocessing.

The disadvantages of this system are the capital cost of the bins. The fact that space in the property must be found for more than one bin, although in many properties this is not a big problem. The residual fraction is only emptied fortnightly (although this has been proved in practice to be quite acceptable, provided a good quality well sealed bin is used).

The advantages are that there is no increase in collection costs (except in the case of the three bin system when it is as small as possible) since the bins are collected in rotation by the same vehicle and crew. Since the bins are clearly colour coded and labelled, good clean separated material is obtained.

Experience has shown that this method is well accepted by the public and is the most cost effective method of achieving the highest possible percentage of recyclable material collected.

The Dual or Multi Bin System is normally used in conjuction with a civic amenity site or "Bring System". Thus the fractions which make the largest contribution to the total tonnage are collected from the house these are normally the organic or compostable fractions and the paper). Smaller fractions such as glass, metal and plastics are then taken to the central amenity site by the residents.

4 EXAMPLES (GERMANY)

There are many examples of these systems in operation with Germany inparticular using a highly developed system of separation of domestic waste into its various fractions. Some of these are:

a) Limburg-Weilburg

This rural authority uses one wheeled bin per house, collected weekly for the disposable fraction. The recyclable fractions are then taken by the residents to the centralised amenity sites using the 3.2 cbm Grumbach containers.

The result is approximately 11% of the total tonnage is recycled.

b) Black Forest

This local authority issues two wheeled bins per house - one grey for the disposable fraction and one green for the dry recyclables. The bins are emptied on alternate weeks, the dry recyclables then being taken to a sorting plant. The result is 23% of total refuse tonnage is recycled.

c) Niederrhein

This local authority has three bins per house. Grey for the disposable fraction, green for dry recyclables which after collection will go to a separation plant and the brown aerated compostainer for the organic fraction. The grey bin and the brown compostainer are emptied on alternate weeks and the green dry recyclables bin is emptied as an additional collection once per month.

The result is more than 58% of the total refuse is recycled.

d) Giessen

In the local authority of Giessen, three bins per house are issued. Grey for the disposable fraction, blue for paper and the brown compostainer for the organic fraction. The grey bin and the compostainer are collected on alternate weeks, while the blue paper bin is collected once per month. Metal, glass and other recyclable fractions are taken down to the central amenity site by the residents and collected in the 3.2 cbm Grumbach containers.

The result is that 42% of the total refuse is recycled without the need for a sorting plant.

c) Goettingen

In this council's area, two bins per house are issued. The grey bin for the disposable fractions and the green compostainer for the organic fraction. These are emptied on alternate weeks. Paper, glass and metal are brought to the central amenity site by the residents and collected in the 3.2 cbm Grumbach Depot Containers.

The result is that 52% of the total refuse is recycled. It is worth noting that 45 kgs of paper per head of population per year are collected here by the "Bring" system, this high figure is accounted for by the fact that the amenity recycling centres are very high density - one per 500 to 1000 head of population.

5 EXAMPLES (UK)

In the UK, systems are not so advanced as in Germany. However, effective systems are being developed some examples of which are:

a) Dundee District Council

This council collects paper in the blue wheeled bin and the organic fraction in the compostainer direct from the house. They also use the 1100 litre wheeled bins to collect recyclable material from centralised amenity sites.

Although these systems are not yet installed throughout the total area the results are very good with more than 30% of the total refuse being recycled in the areas which have the system.

b) Down District Council
(Northern Ireland)

This council has a trial area where three wheeled bins per house are issued - one grey for the disposable fraction, one blue for paper and one brown compostainer for the organic fraction. These are emptied in rotation - the grey bin being emptied fortnightly and the blue and brown bins being emptied monthly on the alternate weeks. The results are very encouraging with about 30% of the total refuse being recycled.

c) Leeds City Council

This council has a two bin system in operation (also some split bins) with the disposable fraction and dry recyclables being collected alternately, this involves the use of a separation plant.

Other council's also conducting trial schemes with Dual or Multi bin collection are:
Kirklees Metropolitan Council
Scunthorpe Borough Council
East Hertfordshire District Council
North Hertfordshire District Council
all these are obtaining encouraging results.

6 CONCLUSIONS

It is clear that systems exist for achieving high levels of recycling quotas.

The systems employed in any given area will depend on such factors as the policital will, the market available for the recycled materials, the finance available and the cost and availability of landfill space as well as the nature of the area.

One thing is certain. The requirement to recycle the maximum possible quantities of our total refuse will continue to become ever more pressing. The systems to achieve this exist and have already been proven in practice.

Why Recycle?, Rainbow (ed.) © 1994 Balkema, Rotterdam, ISBN 90 5410 367 1

A recycling opportunity for waste contractors

R.S. Moore & E.R. Hill
Materials Recycling Management Limited, Harrogate, UK
J.S. Burnett
Aspinwall & Company, London, UK

ABSTRACT: The MRM patented recycling process, which has been assessed independently by Aspinwall, provides a profitable alternative to landfill for the disposal of industrial , commercial and civic amenity site wastes.

1 INTRODUCTION

The main business of private waste contractors is the collection of industrial and commercial waste, together with disposal at landfill sites.

These operators are threatened by increasing regulations on landfill, by reduced availability of landfill sites and by pressures from many directions to switch disposal from landfill to recycling. In order to safeguard their collection business, there is a strong incentive on waste contractors to make a switch in this direction, especially if recycling can be carried out on a profitable commercial basis.

Materials Recycling Management Limited (MRM) have focused their activities on this specific concept and have developed and patented a process for it's implementation. This work, includes feasibility studies, business plan, research and development, ad detailed design package and market research has been partly supported by grants from British Government Departments.

The Department of Trade & Industry, through the Energy Technology Support Unit (ETSU) have commissioned Aspinwall & Company to carry out an independent assessment of the process. Their findings, which are generally favourable, have been published by ETSU in their report no. B1396.

This paper reviews the work carried out by MRM and it's assessment by Aspinwall. It is envisaged that the next stage of development will be the installation of a demonstration plant on a commercial basis, probably in association with a waste contractor.

Although the process is mainly focused on recycling from industrial, commercial and civic amenity wastes, it is also applicable to domestic waste in connection with suitable pre-sort facilities.

2 MRM PROCESS

2.1 *Philosophy*

Recycling of waste for profit is best achieved by suitable selection and broad separation of streams at source, with reasonably large economies of scale, high recovery rates and with plant operations responding to market demands. This represent a change from previous schemes, which have concentrated on a narrow range of recovered materials, resulting in vulnerability to market fluctuations and a low level of recovery. Industrial, commercial and civic amenity wastes, which are not so highly mixed as domestic waste, offer the best prospect for material recovery. Putrescibles which cannot be readily separated are converted into compost and combustibles into a fuel stream.

2.2 *Pre-sorting*

Some wastes, such as those currently delivered to landfill sites, are partly mixed but are worthy of recovery if they are pre-sorted. This process consists of intercepting deliveries prior to disposal, can carrying out primary separating, so that the subsequent operations at the processing centre can be carried out at a lower cost.

2.3 *Integrated Processing*

The MRM process is described in a paper previously presented (Moore & hill 1991), and in the Aspinwall report no. ETSU B1396 (Burnett 1993). The former paper is included as an appendix to the latter report.

The processing centre receives three streams of waste from direct collection, from the pre-sort or from transfer stations. These will have been broadly separated into defined categories with acceptable levels of contamination as part of waste management operations. The waste steams are delivered to the appropriate bays by the following categories:-

A. Compostables and wood (garden wastes, fruit, vegetables, papers, textiles, soil, trees, wooden articles, building timbers, etc.).

B. Mixed secondary organics, glass and metal (paper, plastic, textiles, wood, glass and metal).

C. Paper, plastics and textiles (separate and mixes of these organic materials, with low levels of contamination of other materials).

This demonstrates that selection is on a broad basis and that practical levels of contamination can be readily handled by the processing centre. Colour coding of containers is used to facilitate identification of streams.

Figure 1 overleaf gives a simplified schematic diagram of the process, showing the unit processes involved and the cross overs provided at selected points in the system, so that materials can be diverted from one product stream to another. The individual items of equipment have all been proven in their required service on other plants. The unique nature of the process is in the specific integration of the various streams.

2.4 *Power Generation*

The mixed combustible fraction of the waste provides a fuel stream, which can either be sold as a pellet or burned on site to generate electric power (ideally combined heat and power). The former option provides lower capital and operating costs and a higher return, but involves a more difficult marketing operation. At present there is a fairly strong market demand for pellets.

3 ASPINWALL ASSESSMENT

3.1 *Technical*

Aspinwall engaged Henley Burrowes, who are recognised leaders in the design of waste processing plants, to carry out a technical review of the process. The focus of this review has been on the main processing plant, where materials are segregated and the fuel and compost produced. Previously, Henley Burrowes had prepared a design package for the pre-sort facilities.

The review considered the design of the process plant in the light of more recent experience in the operation of refuse derived fuel (RDF) facilities. There are a few changes required including screening of waste prior to shredding and classification of the mixed stream, automatic eddy current extraction of aluminium and the production of a semi-densified fuel pellet for the power generation option. This minimises storage and transport costs compared to the handling of a fuel floc, whilst at the same time, part densification of the pellets gives considerable energy savings in drying and pelletisation compared to full densification.

Report ETSU B1396 includes Henley and Burrowes flowsheets for the updated process.

3.2 *Economic*

Aspinwall have developed a computer spreadsheet model to assess the economic performance of the plant under various operating conditions, both with the sale of pellets and on-site power generation. Under base case assumptions, the first option produces a very healthy internal rate of return (IRR) of 36%, whilst the second gives a reasonable 22% return. The former option assumes that pellets are sold at a no charge. Thus the return of 36% is considerably enhanced if there is a reasonable market for pellets, which seems to be the case at present.

TABLE 1: Baseline Operating Assumptions

General	
Discount rate	10% per annum
Waste composition	50% commercial/ 50% industrial
Landfill disposal charge	£15.00 per tonne
Project life	20 years
Processing Plant	
Plant capacity	160,000 tpa
Capital costs	£3.7 million
Plant gate fee	£12.00 per tonne
In house power consumption	840 kW
Power Plant	
On site combustion	included
Combustion capacity	48000 t / y
Capital costs	£8.9 million
Calorific value of pellets	18 GJ / tonne
Plant availability	90%
Boiler efficiency	70%
Turbine efficiency	30%
In house power consumption	10% of gen. power
NFFO application	10 years

A series of sensitivity tests have been carried out for both options. These show that the power generation case is more robust under changing

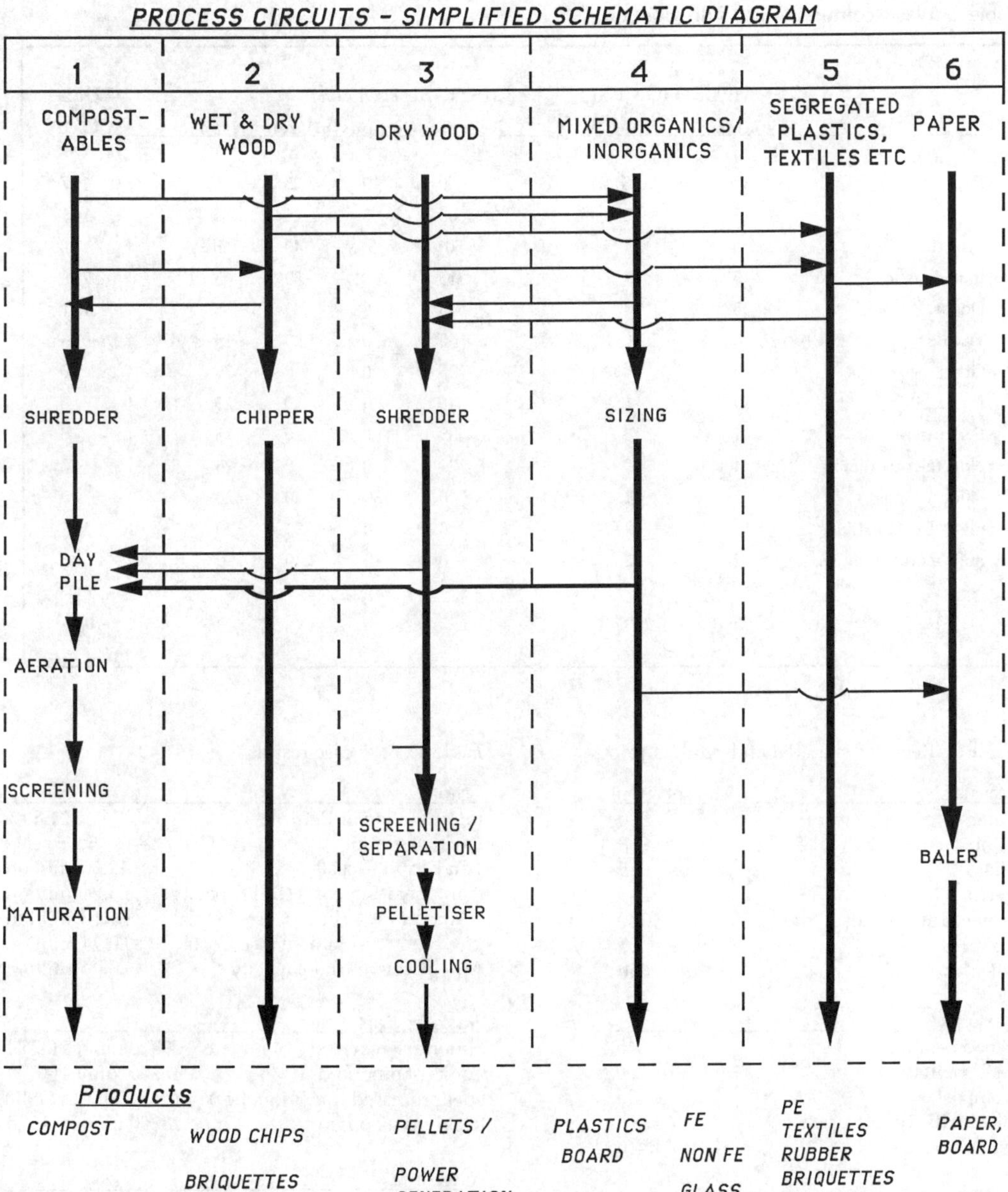

Figure 1.

conditions, such as waste composition, product prices and landfill charges.

This is mainly due to a base case assumption of a Non Fossil Fuel Obligation (NFFO) price of 4.5p/kWh, being available for 10 years, which gives a solid financial base.

These assumptions represent the starting point for analysis, with each of the main variables being subject to sensitivity tests. It is assumed that partial start-up of the process plant will be achieved 6 months after start of construction with full capacity of the process plant and power generation reached after 36 months.

The total capital cost of £12.6 million covers the pre-sort facilities, processing centre and power plant, with all buildings and vehicles leased and farming part of operating costs.

One of the most important assumptions is the composition of the waste, which is poorly defined and is the subject of considerable on-going research.

Table 2. Waste composition and distribution

	COMPOSITION (%)		DISTRIBUTION (%)				FUEL
	Comm.	Ind.	Recycle	Compost	Fuel	Dispose	COMP
Paper	30	8	50	10	25	15	14.7
Card	15	22	65	5	25	5	14.3
Wood	5	25	50°	5ˣ	40	5	0
Film Plastic	5	15	15	0	80	5	24.8
Dense Plastic	3	10	15	0	80	5	16.1
Textiles	3	0	60	0	35	5	1.6
Glass	3	4	25	0	0	75	0
Putrescibles	10	1	0	75	0	25	0
Ferrous Metals	1	5	85	0	0	15	0
Non ferrous metals	4	0	75	0	0	25	0
-40mm Fines	3	1	0	80	0	20	0
misc. Combustibles	16	7	0	0	80	20	28.5
misc. non Combust.	2	2	0	10	0	90	0
TOTAL	*100*	*100*	°chips	ˣmulch			*100.0*

Table 3 Recycled material and product prices

Materials	Market price
Papers	£10
Timber	£18
Glass	£25
Ferrous Metals	£15
Non ferrous metals	£380
Textiles	£5
Plastics	£50
Products	
Wood briquettes	£45
RDF pellets	£0
Compost	£5
Electricity (pool) price	2.5p / kWh
NFFO price	4.5p / kWh

The study is based on recent data from Warren Spring Laboratory, as shown in Table 2.

The base model assumes that four chain grate boilers will be installed each with 16,000 tpa fuel capacity of 18 GJ/tonne calorific value, with three boilers normally on line, burning 48,000 tpa of fuel and giving a power output of 4.91 MW.

Table 3 outlines the assumed unit prices for recycled materials and products, taken from broad averages of published data. Here again, these prices provide a starting point for sensitivity analysis.

Table 4. Base case results

General	
Internal Rate of Return	21.89%
Payback period	5.92 years
Total capital costs	£12.60 million
Ann.operating costs (full capacity)	£3.04 million
including: Labour Costs	£1.10 million
Maintenance Costs	£0.94 million
Ann. revenue (full capacity)	£6.04 million
Processing Plant	
Materials recycled	59720 tonnes (37%)
Compost produced	15200 tonnes (10%)
Fuel processed (inc briquettes)	61280 tonnes (38%)
Residues disposed	23800 tonnes (15%)
Power Plant	
Pellets combusted	51680 tonnes
Ash disposed	5168 tonnes
Rate of combustion	6.56 t / hour
Gross power	1050 kWh / t
Total in house consumption	22%
Net power	816 kWh / t
Export power	5.35 MW

The model has been run using all of the above information as baseline data. Appendix 2 of report ETSU B1396 gives a full printout of assumptions, input and output, which is summarised in Table 4 .

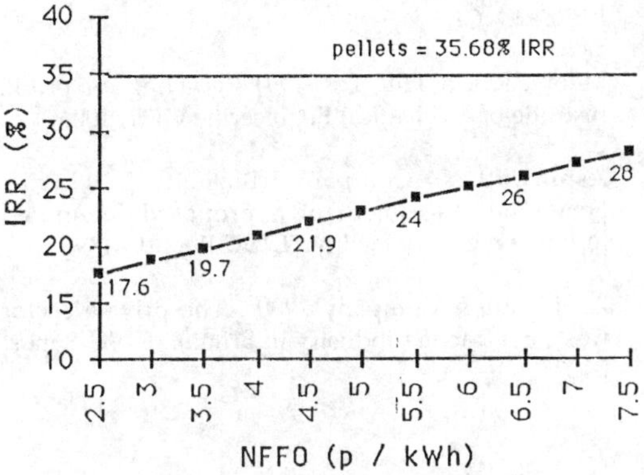

Figure 2. Sensitivity to NFFO strike price

Table 5: Summary of Sensitivity Studies

Variable	Value	IRR % Pellets	IRR % Power
Pellets to landfill - % of total	0	36	(NA)
	60	22	
	100	14	
Waste composition - % commercial waste of total waste	0	28	18
	50	36	22
	100	44	26
Land fill charges - £/tonne	5	44	25
	15	36	22
	25	28	19
Calorific value - GJ/tonne	15	36	20
	18	36	22
	22	36	25
Gate fee - £/tonne	8	21	16
	12	36	22
	20	68	33
Product prices - % change	(20)	27	19
	0	36	22
	20	47	25
Capital costs - £m	3.0/10.1	43	29
	3.7/12.6	36	22
	4.6/15.8	30	17
Labour costs - £m	0.9/1.1	40	26
	1.1/1.4	36	22
	1.4/1.7	31	18

These results of the sensitivity analysis in report ETSU B1396 are given in ten charts, showing the change in Internal Rate of Return for percentage change in selected variables. One of these charts is included in this paper as Figure 2 showing the method of presentation used in the study. The other charts are summarised in Table 5, showing the base case values (36% IRR for pellets and 22% for power) together with minimum and maximum cases for the range of variables studied.

The sensitivity studies show that the rate of return is most influenced by the gate fee and secondly by product prices, which is in line with the general perception of the economics of recycling. The pellet disposal study assumes that in all cases there is no revenue from the sale of pellets. Thus, if increasing quantities of pellets have to be disposed to landfill, there is a reducing rate of return.

The landfill charges study reflects the increasing costs for disposal of residues with increasing landfill charges. In practice, increasing landfill charges will be accompanied by increasing gate fees, resulting in increasing rates of return when these two variables are taken in tandem.

The overall sensitivity study gives the values in Table 6.

Table 6. Overall Sensitivity Study

	IRR % Pellets	IRR % Power
Maximum return	68	33
Minimum return	14	16
Base case	36	22
Average of maxima	45	27
Average of minima	27	18
Arithmetic average	36	22

The two options given offer a choice between a low risk, moderate return design for power generation and a higher risk, high return design for marketing of pellets. The latter risk can be reduced if reasonable revenue can be obtained for the sale of pellets, since all cases assume no revenue from this source. The characteristics need to be tested in more detail against local conditions for a specific project.

3.3 *Location and Marketing*

The first factor affecting the location of the recycling plant is an adequate supply of suitable feedstock, most likely available from commercial, industrial non-hazardous and civic amenity wastes. Major waste contractors will most certainly have existing contracts for the collection and disposal of these waste streams. Data has been collected on the composition of these streams, and further research is being carried out by Warren Spring Laboratory and Aspinwall amongst others.

The study assumes that:-

- The quantity of commercial and non hazardous industrial waste generated nationally is approximately equal to the national tonnage of household wastes.
- This tonnage is distributed between regions in proportion to the numbers of persons employed in manufacturing in each region.

This enables waste arisings to be determined on a regional basis, using Chartered Institute of Public Finance and Accountancy (CIPFA) statistics on household wastes and government statistics on Regional Trends in Employment.

The second factor of key importance to long term viability is the availability and location of suitable markets. MRM and Aspinwall have researched these on a company basis and they are identified by name and map location in an appendix to report ETSU B1396. Since transport costs are an important factor in waste handling, a simple scoring system has been used to range counties in terms of their accessibility to recycling facilities.

A further scoring systems has been developed to reflect the variation in landfill charges in different parts of the country. The overall locational analysis shows the most favourable areas in England to be:-

- Greater London
- West Yorkshire
- West Midlands
- Greater Manchester

Followed by:-

- Staffordshire
- Derbyshire
- Hampshire

4. CONCLUSIONS

The MRM process provides an attractive opportunity for waste contractors to move into the recycling of their existing waste collections on a commercial basis. This secures a safeguard to their collection business, by providing an alternative disposal to landfill, whilst at the same time being more acceptable than incineration. The Aspinwall study shows the robustness of the process and identifies those factors which are of key importance.

REFERENCES

Moore, R.S. & Hill, E.R. 1991 Recycling for profit. Institute of Mechanical Engineering C421/1006

Aspinwall & Company (Burnett, J.S.) 1993. Economic assessment of a proposed integrated resource recovery facility. ETSU Report B1396

Aspinwall & Company 1990. The private sector waste contracting industry in Britain. ETSU Report B1238

Why Recycle?, Rainbow (ed.) © 1994 Balkema, Rotterdam, ISBN 90 5410 367 1

Rural Mini Recycling Centres

C.J.Carter & A.B.Cavey
Arup Environmental, Ove Arup & Partners, London, UK

ABSTRACT: The Government is committed to increasing the level of recycling to 25% of total household waste by the year 2000. If this target is to be met the establishment of a comprehensive, countrywide network of recycling schemes, both in urban and rural areas, is required. The authors have undertaken a study of "Rural Mini Recycling Centres" for the Department of the Environment (Yorkshire and Humberside Region). The study revealed there to be a great variability in the design, operation and location of rural mini recycling centres across the country with no single definitive method by which to establish a new scheme, with localised factors having an important influence over decisions made on all aspects of design, location and operation. After briefly reviewing the background to recycling in rural areas, this paper addresses the following specific issues:-

- Choice of regional and site specific location;
- Site presentation and choice of containers;
- Methods of funding.

1.0 INTRODUCTION

In 1990 the Environmental Protection Act introduced a waste management framework which identifies the individual local authorities as responsible for investigating appropriate methods to facilitate recycling, and for compiling a waste recycling plan. In Part II, Section 50 of the Act it is required that where 'reasonably practical' priority is to be given to the recycling of waste, and Section 52 identifies the arrangements for payments for expenditure saved by recycling. In support of this policy the Department of the Environment published Waste Management Paper Number 28 in 1991 to provide guidance to local authorities on recycling (Department of the Environment, 1991). Recycling has been formally defined by the Government as "the collection and separation of materials from waste and subsequent processing to produce marketable products" (Department of the Environment, 1991).

The Government has set the target of recycling half of all recyclable household waste by the year 2000. If this is to be achieved comprehensive recycling schemes in both urban and rural areas need to be developed. At the time of the study only an estimated 2.6% of the total household waste stream handled by the local authorities was being recycled (Friends of the Earth, 1991(a)). The majority of recycling schemes implemented to date have been concentrated in the larger cities and conurbations. However, with the existence of often well established local community networks, there is the potential within rural communities to provide a successful base for the establishment of recycling schemes.

A recent survey showed that 81% of the population in the UK are keen to recycle (Forshaw et al, 1990). However, experience has shown that response rates to recycling schemes are generally low, with around only 40% of the community participating. The main disincentive to recycling appears to be lack of convenient facilities. A survey of 1200 people undertaken by Save Waste and Prosper Ltd in the north of England showed that of non-recyclers, 47% would participate if there were facilities in the immediate neighbourhood. Other factors which have been shown to influence the degree of participation include:

- how well the scheme is publicised;
- how 'user friendly' the scheme is;
- socio economic circumstances;

These factors apply equally to both urban and rural schemes and indicate the potential scope for increasing the success of recycling schemes if they are designed appropriately.

This paper is based upon a report compiled by Ove Arup and Partners on behalf of the Department of the Environment (Yorkshire and Humberside Region), with the objective of reviewing the current operation of mini recycling centres, (MRCs), in rural areas and to make recommendations on the optimisation of their design, location and operation. For the purposes of this study 'rural' was defined as any village or hamlet with a population of less than 2500 people.

MRCs are centralised collection points consisting of a number of containers to which the community bring, and deposit, recyclable materials. They are relatively cheap to establish as some transport costs are borne by the public.

This project involved a literature review, discussions with appropriate organisations, and a visit to a local authority which was in the process of planning and designing a pilot network of MRCs. In addition a brief questionnaire was sent to a small number of merchants involved in buying and selling reclaimable material in order to obtain feedback on their operational requirements for MRCs. This paper addresses a few of the aspects covered in the course of the study including the choice of location, site presentation, choice of containers and methods of funding for MRCs. The full report, (Arup Environmental 1993), also discusses the choice of materials for collection, space and access requirements, pilot schemes and public education programmes, and contains details on three rural schemes which were considered to be operating successfully. The potential criteria by which schemes may be defined as successful were also discussed.

2.0 BACKGROUND TO RECYCLING IN RURAL AREAS

There are currently a wide range of recycling schemes operating in rural areas ranging in scale from provision of simple recycling directories, single material collection points, (eg bottle banks), and some kerbside collections. However, it has been demonstrated in a recent report (Shankland Cox and Cremer and Warner, 1992), that in rural areas with a low density, dispersed population, the most effective method of recycling is by the establishment of 'bring' systems of collection, and in particular MRCs. MRCs do not generally achieve as high collection rates as kerbside collection systems, with recovery rates generally in the range of 15-20% of the total household waste (Friends of the Earth 1991(b)). Table 1 shows typical quantities of material collected by each type of system in rural areas.

A number of advantages and disadvantages are cited for recycling in rural areas, some of which are considered in Table 2. Although the disadvantages appear to outweigh the advantages in absolute terms the strong community associated with some rural communities can more than compensate for these shortcomings in some instances.

Table 1 : Typical Quantities of Materials Collected by Rural Schemes in tonnes (Arup Environmental 1993)

Scheme	Total Population Served	Paper	Cans	Glass
Presteigne† Wales 1990/91	2,000	80	0.33	62
Test Valley† Borough 1990/91	106,000	-	2.1	17.95
Ivybridge† District Devon 1989/90	8,520	44.34	-	-
*Mochard Bishop, Devon 1990/91	930	26	13	13
*Wye Cycle 1991	2,000	78	8	52

† MRC schemes. Figures as average for a single MRC.

* Kerbside collection. Figures estimated from weekly collection figures

The profitability of recycling schemes, particularly in rural areas, is becoming increasingly marginal. Market prices for the collected materials will ultimately determine the feasibility of a scheme. Unfortunately these are traditionally volatile and vary greatly for different materials and geographic location. Generally aluminium cans and glass cullet are more profitable than waste paper which can be notoriously difficult to sell. However, there is some fear that the price of glass cullet bought by the recently formed British Glass Recycling Company, one of the major buyers in the industry, may fall in the near future (Anon, 1993(a)). In order to overcome such problems in the medium term, 2-3 year contracts need to be secured with the merchants prior to

Table 2 Some Advantages and Disadvantages Associated with Recycling in Rural Areas

Advantages	Disadvantages
1.It is generally perceived that rural communities tend to have stable populations and strong community spirit.	1.Small, dispersed population and therefore small absolute amounts of waste. Rural populations generally produce less waste per head of population.
2.Security and vandalism of containers is generally less of a problem than in urban areas.	2.High transport costs due to relatively large distances involved.
3.More scope for reducing waste by other complementary means such as composting.	3.Apparent lack of suitable sites for location of MRCs.
4.Potentially large reductions in cost of traditional rural refuse collection schemes.	4.High capital costs of scheme start up.
5.Public education and awareness of schemes may be more effective.	5.Lack of financial incentive - recycling credits potentially lower in rural than urban areas.
6. --	6. Lack of competition between merchants, therefore often low prices paid for materials.

Table 3 - Typical Prices for Recovered Materials 1993 (Anon, 1993(b))

Material	**Price (£ per tonne)**
Paper	0-120
Steel cans	10-20
Plastic	0-25
Aluminium	600-700
Textiles	98-154

commencement of operations. Typical prices for materials for 1993 are given in Table 3

Management of rural recycling schemes appears to be undertaken either by a local voluntary group, by the local authority or by some form of joint arrangement between the two. In response to recent legislation most local authorities are now operating some form of recycling scheme, with some appointing specific Recycling Officers to oversee their operation.

The following sections of this paper consider three important aspects of the management of MRCs in rural areas than if undertaken appropriately can contribute considerably to the success of the scheme.

3.0 LOCATION OF MRCs

Good location is a fundamental requirement if a scheme is to be well supported. Regional location will need to be considered in the context of each specific region, and locally operating factors including:

- Population catchment;
- Quantities and types of material generated;
- Geographic and demographic dispersion patterns;
- Existing waste collection routes;
- Local support.

Little work has been done on the optimum population density required to support a site. Higher densities of MRC's per community are common in the rest of Europe than those currently found in the UK, with correspondingly higher recovery rates. Most schemes here tend to be designed on the basis of one MRC per 400-750 population (based on the use of four 1100 litre wheeled bins), however the range is wide from 100-2000 population per scheme. Specific location then tends to be either in a single central village, as a network of sites on a linear basis, or in the nearest market town. The main requirement is that the site is located on a route, or close to a facility/structure, well frequented by the local population.

Alternatively some organisations have chosen to distribute resources as a direct result of public support for the schemes, rather than on a population basis, eg Ryedale District Council and City of Dundee District Council. Hence, it is possible for a relatively small population to support a scheme more successfully than a larger village which may lack enthusiasm for the project.

Site specific location within a village or town can be a sensitive decision. It will often require prolonged discussion and negotiation with local groups and residents. Ideally the site will comply with the following criteria:-

(i) The site must be readily accessible;
(ii) No special trip is required by the householder to utilize the facility;
(iii) It will not cause a nuisance to the local community;
(iv) Adequate space is available to service the MRC;
(v) No planning problems are associated with the site;
(vi) Site security will not be a problem;
(vii) It is preferably a site on sufficient hard standing.

Generally a number of locations appear to be favoured including pub, village hall and council car parks; village greens; school playgrounds; delivery bays at local shops and garage forecourts.

Pub car parks are particularly favoured since the MRC can attract large volumes of materials eg. cans from the pub itself, and the sites tend to be hard paved with plenty of access space. Sites on hard standing areas are particularly important for the winter months when a well frequented site can become muddy and consequently unattractive to use.

Sites which already operate with a single bottle bank or skip may be developed, as these have already been chosen for their various attributes and are probably well known to the local community. Potential concerns over noise, litter and visual intrusion can usually be amicably overcome by open discussions with the concerned parties, and if required by compromise with the provision of litter bins, restriction on operating hours or the installation of soundproofed containers. Discussions with local bodies such as the Parish Council can avoid such problems as they are usually acquainted with potential sites which are well frequented and unlikely to cause a disturbance or concerns to the local community.

Planning permission for an MRC may be required under the Town and County Planning Act 1971, although generally local authorities appear to be more flexible, particularly if wheeled bins are used.

Technically these are not considered permanent. Application for planning permission can be a lengthy process lasting up to 6 months, although without it theoretically a project may only proceed for 28 days in any 12 month period.

4.0 CHOICE OF CONTAINERS AND SITE PRESENTATION

A wide range of containers are available for use at MRCs. These are generally classified by the different types of equipment needed to service them, i.e. wheeled bins, skips, 'roll-on roll-off' containers, igloos, and modular bottle and can banks, as described in Table 4. The most popular choice appears to be the use of small volume wheeled bins, in particular 1100 litre capacity bins, although these may not be suitable for bulky material such as paper. The precise choice of size or type of collection container will depend on:

- Character of the material collected;
- Quantities generated and frequency of collection.
  Minimising transport costs requires maximising the amount collected per load;
- Ground area available;
- Aesthetic requirements;
- Hygiene requirements;
- Costs;
- Security and durability.

Table 5 indicates typical costs for equipment at the time of the study.

There is a move towards visual uniformity of container size, type and colour in an attempt to optimise site appearance, although in some schemes different coloured bins are utilized for different materials. In some instances colour coding is undertaken to identify containers for specific materials, for example green for green glass, brown for brown glass and yellow for cans.

Generally dull colours such as green, brown or grey appear to be favoured as these are more acceptable in aesthetic terms as they blend in well with the surrounding vegetation and buildings.

Some ancillary equipment may be required at some sites in order to maximise the schemes efficiency or ensure the site is kept tidy, for example, magnets for aluminium can banks, can crushers and litter bins. Some sites have experienced considerable problems with fly tipping and litter if suitable disposal facilities have not been provided.

Signing is a design factor that is important to the success of the scheme. Due to the high costs involved some operators argue that it is not cost effective for a small scheme, particularly when the MRC is located in a prominent location. Although this may be the case in a few locations, for most sites in larger villages and towns, particularly those which have been screened or landscaped, there is a need for the clear and appropriate signage of sites. Signs must be clear and effective at presenting the required information regarding the nature of the site, preferably using standardised colours or logos to make them instantly recognisable. Whichever system of signage is adopted it is important that the following information be provided in some form;

(i) information on types of materials collected;
(ii) operators name;
(iii) hours of operation;
(iv) contact name for enquiries.

The most comprehensive signing system discovered in the course of the study was that used by Merseyside Waste Disposal Authority who used road signposts, entrance signs (giving details of opening hours, materials collected and contact name), free standing pedestals, and container signing. The signs were made from painted hardboard with a printed vinyl overlay which was considered the most effective and suitable construction.

Landscaping of MRCs does not appear to be undertaken on any scale, possibly due to cost constraints, but it may need to be considered when the location is sensitive or close to housing. In such cases the most common form of screening is larch lap or vertical closed fencing. Only in a few cases have trees or shrubs been planted.

A complementary report by Neap Environmental Consultancy Ltd (DoE, 1992) looked at the presentational design and signing of sites in some detail.

Table 4 Main Types of Container Used in MRCs. (LARAC, 1987)

Container	**Size**	**Comments**
Traditional Bottle Bank	c.12m^3	Uses skip moving vehicles. Hard to vandalize irreparably. Noisy if not soundproofed.
Duobank	c.7m^3	Has to be used in pairs on sites with slower filling rates.
Igloo	0.9-2.5m^3	Have tubes to stop noise and glass removal. Easily vandalized.
Durabank	4m^3	Easily burnt or damaged beyond repair.
Paladin	1m^3	Can be used with standard lifting equipment. Low payload.
Eurobin	1.1m^3	Wheeled, therefore easy access but need locking to prevent freewheeling.

Table 5 Typical Costs of Equipment

Type of Container	Approx. Cost New
1100 litre Wheelie Bin	£250-300
12m^3 bottle bank	£1,100
Igloo	£350
12m^3 skip type compaction container	£7-10,000
Can sorter and baler	£6-10,000
Small baling machines	£600-2,000
Waste Roller Baler	£5-6,000
Glass Crusher	£5,000

(Source: Various Suppliers 1992)

5.0 METHODS OF FUNDING

The capital investment required to set up and operate MRCs can be considerable and quite often this is not offset totally by the revenue from the sale of the recovered materials.
Some schemes have experienced cash flow difficulties due to delays in payments made by merchants for recovered materials. Consequently it is advisable to compile a detailed financial plan during the design stage of the scheme in order to ensure that such contingencies are planned for.

Theoretically some funding is made available to the local authorities by means of 'recycling credits', which are meant to represent savings made in the collection and disposal of waste by diverting a given volume to recycling schemes. These credits were introduced under Section 52 of the Environmental Protection Act and are paid by the Waste Disposal Authority to the Waste Collection Authority, or to other organisations which are actively recycling. The monetary value of these credits are considerably lower for rural areas than for urban areas, eg for 1993/94 £4.62 compared to £16.95 per tonne for inner London (Anon, 1993(c)). The Government has promised to double these values by April 1994 in order to take into account the marginal costs of waste disposal (Anon, 1993(c)), but the discrepancy between the payments in the two areas will remain. This means that recycling credits are not likely to provide any major financial support for MRCs in rural areas.

Fund-raising events are the most obvious

method by which small, particularly rural, communities are able to raise resources for recycling schemes. However, there are a number of alternative sources of funding from external bodies including grants and loans. At the time of the study organisations offering financial support included UK 2000 Projects Grant Fund, Shell Better Britain Campaign and various local authority and government awards. The size of grants varied from ten pounds to thousands of pounds.

A popular alternative source of funding appears to be sponsorship from local companies, businesses and newspapers. Businesses presently appear to be comparatively keen to become associated with 'green' community projects in return for public recognition of their support. While recycling schemes remain relatively novel, sponsors may receive the publicity and profile they seek. However, this situation may change as recycling schemes become more widespread.

6.0 CONCLUSIONS

1. Recycling can be financially viable in rural areas if the specific conditions of the locality are taken into account, and the recycling strategy designed accordingly. There is no single definitive method by which to design and operate an MRC.

2. Most rural schemes appear to be utilising low volume wheeled containers, typically in groups of four or five, one for each type of material collected.

3. Loads should be maximised by optimising collection frequency if recycling is to be efficient. Containers should be selected taking into account the quantity of materials likely to be generated, equipment available to service them and space and capital available.

4. The choice of site is very important and should be conveniently positioned such that householders are not required to make a separate journey to use them.

5. The design of the MRC should be as simple and 'user friendly' as possible. Site presentation is important with containers typically in dull colours such as brown's greens and greys which are less visually intrusive. All sites should be clean, tidy and clearly signed.

6. Wherever possible secure, long term contracts should be negotiated with merchants before collection is commenced, and all possible sources of funding investigated. A detailed financial plan should be drawn up during the planning stage.

7. Although the success of a scheme may be readily quantified in terms of its profitability, other, unquantifiable benefits, such as the development of a community focus, or gratification of participating individuals, may overall be equally if not more important criteria in the assessment of the overall success of the scheme.

ACKNOWLEDGEMENTS

Arup Environmental would like to thank the Department of the Environment (Yorkshire and Humberside Region) who have kindly given their permission for the publication of this paper.

REFERENCES

Anon, 1993(a), Price Cut Puts Glass Recycling in Jeopardy says Councils, Ends No 224, 12-13.

Anon, 1993(b), Recyclers Move the Waste Mountain, Local Authority Waste and Environment, **1**, (5), 22-24.

Anon, 1993(c), Recycling Payments Increased, Ends Report No 218, 34.

Arup Environmental, 1993, Rural Mini Recycling Centres, HMSO.

Department of the Environment, 1991, Waste Management Paper No 28, HMSO London.

Forshaw, J., Hays, A. and Wright, C., 1990, Fashionable Waste. The Make-up of a Recycler, Save Waste and Prosper Ltd, Leeds.

Friends of the Earth, 1991 (a), A Survey of Local Authority Recycling Schemes in England and Wales, Friends of the Earth Trust Ltd, London.

Friends of the Earth, 1991 (b), Recycling Officers Handbook, Friends of the Earth Guide for Local Authority Officers, Friends of the Earth Trust Ltd, London.

Local Authority Recycling Advisory Committee, 1987, Expanding Glass Recycling - A LARAC Guide.

NEAP Environmental Consultancy Ltd, 1992, MRC's - Practical Guidelines for Improving Design and Community Awareness, Department of the Environment, Yorkshire and Humberside Region.

Shankland Cox and Cremer and Warner 1990, Recycling in Rural Areas, A Feasibility Study, Department of the Environment, HMSO, London.

Why Recycle?, Rainbow (ed.) © 1994 Balkema, Rotterdam, ISBN 90 5410 367 1

Author index